VIETNAM

On My Mind

VIETNAM
ON MY MIND

Stories by
Robert Brundrett

Juli -
A pleasure meeting you!
Good luck in your career!
RWBrundrett
4-19-2014

The Wooster Book Company
Wooster Ohio
2013

The Wooster Book Company
where minds and imaginations meet

Printed and bound in the United States of America.

ISBN: *978-1-59098-739-1*

The Wooster Book Company
205 West Liberty Street
Wooster, Ohio • 44691
www.woosterbook.com

This book is printed on acid-free paper comprising at least 30% post-consumer recycled fiber.

To Linda

Table of Contents

VIETNAM
On My Mind

Prologue

Vietnam has been on my mind. It's been forty years since I spent a year over there during the war, and lately I've found myself thinking more and more about that time. As an advisor there, I worked with an interesting assortment of people. I still think of some of them from time to time, and often find myself wondering how their lives turned out. I decided to develop some of my recollections into stories for a book. The stories are loosely based on actual events, but all names are fictitious.

This is not a collection of war stories. Nor is it about the significant events of Vietnam during that year; it's about the everyday events, everyday people, and the places that have remained with me all these years. When I think of Vietnam, my mind wanders to a Chinese man and his three sons who drove around in a white Datsun pick-up, or a fearless US Navy salt with a rubberized face who liked to sing sea chanteys, or an old Vietnamese man after losing his son in a fire fight, or a young Vietnamese girl who carried cement bags on her shoulders. It is strange but it's the insignificant things that the mind seems to retain while it glosses over the major ones. The insignificant becomes the significant with the passage of time.

I joined the Navy in 1969 after working a year at General Motors as an engineer following my graduation from Ohio State. When I got orders to Vietnam in May of 1971, I had a bad attitude about the whole thing. My wife Linda and I had been married for a matter of

San Francisco Bay was good Navy duty.

just months. We were stationed at a Navy base on San Francisco Bay. It was a great job at an outstanding location; we had a nice apartment and wonderful Navy friends. I was a Lieutenant Junior Grade (LTJG) with not too much responsibility, and most weekends and evenings were free. And who could fault San Francisco? We had it made. I figured I'd be there for another year before my regular discharge. My greatest concern was planning my career as chief engineer at General Motors or NASA after my Navy stint ... but I could do that in the upcoming year ... plenty of time.

My first reaction when I got my orders was: What? Why now? What's the point? Everyone knew in mid-1971 that the long Vietnam War was nearly over. And who wanted to be the last one in Vietnam? We all knew how it would end: the American war machine would wind down and we'd just quietly leave. That process had already started. Also, before I shipped out, I would be required to participate in six months of advisor training, and wouldn't even arrive in Vietnam until January '72. Would there even *be* a war by the time I got there? Everyone was tired of Vietnam.

The bad attitude continued for a month while I tried to think of ways to avoid Vietnam. By "avoid" I don't mean moving to Canada; I mean trying to convince my detailer that I'd be more valuable to the Navy Department using my engineering degree right here in the good old USA. Unfortunately, the US Navy didn't share that opinion and reality gradually set in. I begrudgingly started training. But I found the training to be a challenge, and took an interest right away. As an advisor, I was required to take courses in Asian culture, Vietnamese language, and military survival training in case of capture. I was not enthralled by the military training in the mountains outside San Diego, but I became fascinated with the language and culture of Vietnam.

When I got to Vietnam, my interest in the land and its people continued to grow. Today, most of the military memories have faded, but the people I knew and worked with are as clear in my mind today as they ever were. And that goes for Americans I knew as well as Vietnamese; and not just those with whom I worked closely on a daily basis. Many a chance meeting has stayed with me all these years. That's what these stories are about. They're about a few of the many interesting people I knew in Vietnam in 1972 and early 1973 as the American portion of this war was ending.

I was officially part of the Naval Advisory Group, Republic of Vietnam and was one of twelve Civil Engineer Corps officers that were scattered along the coastlines and rivers at Vietnamese Navy bases to support the public works, construction projects, and defense of the bases. We were assigned to "counterparts," Vietnamese naval officers of like rank to ourselves, with whom we worked and often traveled throughout South Vietnam. My job, in particular, required me to travel a good deal across the countryside of Vietnam, and some of the stories emanate from incidents I experienced during those trips. Though I didn't plan it this way or know it at the time, my experience in Vietnam gave me a front row seat to arguably the largest and most defining event of the baby-boomer generation. And for that I am grateful.

George Loved Fords

Seaman George Belden was the nervous type. He was a friendly guy, but not the kind you would want to rely on if something went wrong. And things were always going wrong in Vietnam. Sometimes they were quite humorous. I was with George twice when things went wrong. This is one of those times.

George had had a year of college, but had decided the academic life was not for him, and since he had a low draft number, he had enlisted in the Navy to avoid the Army and Vietnam. Obviously his plan wasn't working very well, and he ended up at Qui Nhon, a coastal patrol boat base along the central coast of South Vietnam.

George and I always made the trip to Phu Cat together. We went once every two weeks to retrieve cash to pay the native workers on our base. As engineering advisor on base, paymaster was one of my collateral duties, so I went to Phu Cat to get the cash. I guess George went to drive and protect me with his M16. Anyway, he was good company and two heads were always better than one. We always took the Ford jeep. In Vietnam, most of the jeeps were made by Ford, not Jeep. If you had a jeep made by Jeep, though, it was a real jeep; if not, you had a Ford jeep. You could tell them apart by the louvers on the radiator grill. The Ford's were horizontal. But the real Jeep's were up and down, just like they still are today.

George and I always took the Ford Jeep.

One overcast morning in March '72, George and I headed out for Phu Cat Air Base once again. George had signed out the Ford jeep and I met him at the ferry pier. Our little base was on the Qui Nhon peninsula and it was necessary to get over to Qui Nhon City to go anywhere by vehicle, so we took the ferry which ran over and back four times a day.

George was the designated driver, but for the last three trips I'd been driving the jeep onto the ferry. This was because George said he was a city boy and never got the feel of how to bounce the jeep onto the Mike boat. The Mike boat had a big steel gate that slammed down on the concrete beach and became a ramp to drive aboard. "It was a matter of momentum and appropriate clutch work," the ferry chief had tried to explain to George. The chief was raised on a farm, and said it was akin to bouncing a tractor into a concrete feed lot. Since I was also raised on a farm and familiar with the "bouncing," as it were, I usually volunteered for the ritual. This kept both the chief and George happy. Once the jeep was parked on the boat, though, George would drive the remainder of the trip.

By the time we crossed the bay, the sun had burned through the low clouds. George climbed into the driver's seat and we took off through the city and then onto Highway 19 toward Phu Cat. It was mid-morning, the sun was out, temperatures were pleasant and very few vehicles were on the road, so we went breezing along … it was going to be a great day. Our weapons consisted of George's M16 and my .45 pistol and maybe a few extra bullets. Binh Dinh Province was supposed to be pacified. It was sometimes referred to as a fragile pacification, however.

The countryside was very green but the tree and large growth vegetation was low in height and a good distance from the road. We could see large expanses of crops, mostly rice, which gave the appearance of great lawns on either side of the road. We could see for miles ahead as the road curved and wound its way on the flat countryside north to Phu Cat Air Base. Occasionally we would pass an old villa set far back, perhaps one-half mile from the road. Built by a French planter perhaps a hundred years prior, it was probably now used as a business center for the rice farms by a well-to-do Vietnamese landowner.

As we drove along, I saw George beginning to look around more frequently on both sides of the road and in the rear mirror. He seemed tense as he remarked, "You know, we haven't seen a vehicle or in fact another person now for about fifteen minutes." I said nothing, but took note myself. As part of the Engineer Corps, I also noted the jeep was beginning to sound a bit rough, but didn't share my thoughts with George.

We motored past a rather large plantation of rubber trees, neatly lined up, with a great campus of old tile-roofed processing buildings as we continued on through the tranquil rural countryside. The road was narrow by US standards, but in decent shape and still no traffic. George kept the jeep at a steady clip of about 45 mph, and was beginning to relax. I could hear him humming, *"Bye, Bye, Miss American Pie ..."*

I had made it a point to keep an ear to the vehicle, and after another fifteen minutes of driving realized the roughness had turned into a definite clunking sound. A well-defined clunk was now getting worse; I tried to estimate how far it was to Phu Cat, and our chances of making it there without breaking down. I decided we'd never make it, and thought it best to stop and check the vehicle. But I knew this was going to make George nervous.

I was looking for a good place to pull over when all the trouble started. Just as George was mouthing, *" ... drove my Chevy to the levee ...,"* a large bang shuddered through the vehicle and the jeep coasted to a stop despite George's pumping the gas and urging, "Go you son of bitch! ... Go! ... Please!" George had pulled over onto the grass as the vehicle came to a stop. The mood changed immediately. The engine was still running; I told George to shut it down. After that it became real quiet, the only sound was a light breeze. I glanced around and could see we were on a flat section of road with no trees or cover in any direction except for the slight ditch and the green fields of grain on both sides of the road, waving lazily in the breeze. Up ahead I could see there was a small bridge less than an eighth mile away, and then the road turned left and out of sight behind a stretch of brush and small trees.

I looked over at George and saw only anxiety building. His hands had a death grip on the wheel; and so, trying to act nonchalant, I said, "Let's take a look-see, maybe we can get this critter rolling again." George fumbled for his M16 and we walked up onto the road to scan the area. "I don't like it ... it's too quiet. Why in hell does this crap always happen to me!" George said as he kept backing around in a circle checking the surrounding area. George was a tall boy, maybe six-three, and he had a kind of scrubbed clean yet tousled look, like his mother would want to take him by the shoulders, and straighten out his collar before wheeling him around and pushing him off to school. He looked as though he'd been pushed up on the road.

It really did seem peaceful, though, with the rice fields waving and a warm pacific breeze, and we stood there for a moment to take it all in. We could have been standing in a wheat field in Ohio for that matter; or maybe we allowed ourselves the luxury of thinking we were for a few moments. There was no one or anything around. The only distinct sound I remember was George's boots on the pavement as he made his backward circles.

I popped the hood on the jeep, and as I bent over to check the engine, the silence was broken by a deafening explosion across the road up ahead. I remember it threw water and mud debris high into the air, some of which landed on the road. I dived for the ditch and George, standing opposite the jeep, dived under the jeep. Then, nothing, just pure quiet again. George mumbled, "Crap! I knew this would happen to me! I just knew it!" George was on his stomach with the M16 pointing out at the small trees up ahead. I crawled up to the jeep to talk to George. We both agreed that whatever it was, it hit about three-hundred feet up the road near the small trees.

We lay there quietly thinking about our alternatives and fate when the second explosion went off, which seemed closer. "The bastards are walking mortars toward us!" George cried, "I just knew this crap would happen to me; my mother told me I would make a pitiful soldier!" I could see George moving around and was a little worried he was going to make a run for it in the rice, but then with a voice as controlled and as proud as could be, George exclaimed, "Hey Lieutenant, I think I just found our jeep problem." And he had; he was lying next to a broken left rear axle shaft. The CV joint had failed at the wheel and the axle was hanging down from the differential housing.

We both crawled in for a better look and pondered the situation, anticipating the next explosion at any time. Repairing or replacing the part was out of the question, and waiting around for a friendly ride in our situation didn't seem appealing either. Then, in a classic moment of stating-the-obvious, we both said, "Hey, this is a four-wheel-drive vehicle!" and we were using only two wheels, the rear wheels. I remembered in the Ford jeeps, you could go all four-wheel-drive, or either rear or front drive separately, so if we disconnected the rear and drove in only front drive we should be good. George said, "If we get back alive I'm buying a Ford car, good old Fords … I love Fords." All we needed was something to tie the hanging axle to the jeep's frame so it wouldn't drag or catch on something.

George reached up over the side of the jeep with his long arms and rummaged long enough to pull down a roll of communication wire. It had been several minutes since the last explosion, and we were sweating and working like crazy in an effort to secure the axle, when we both stopped abruptly. George whispered, "Voices, I hear voices, son of a bitch, I hear voices!" "Vietnamese," I said, "sounds like they're near that little bridge, on the left side of the road." We held our breath and listened. As the breeze shifted slightly, we caught the sounds again. "From the direction of those sounds, I'd bet they're a ways back from the road," I whispered again. "We're not that far from Phu Cat; we may be able to get out of here straight up the road if we get this axle tied up quick."

We finished securing the axle. George crawled out from under the jeep and came up the driver's side while I slipped into the passenger's. I messed with the gear shifters until I was sure we had only front drive selected, then in one continuous movement, George hit the starter and gunned the engine. The jeep sprang to life and jumped up on the road with the front wheels throwing gravel and rubber. He pinned the pedal to the floorboard as the little Ford accelerated quickly up the road, while we both hunkered down in the seats as far as possible with weapons drawn.

As we approached the bridge the third and largest blast occurred with a great deafening explosion while heaving large amounts of mud, water, and grass into the air with some of it splashing on the windshield. A piece of debris that looked like a fish's head bounced on the hood of the jeep and wedged on the windshield. George held the wheel in a fierce grip as he yelled "What the heck is that? Mother of God, please let me live!" The adrenalin kept his foot solid on the accelerator and the jeep gained speed as it entered the bridge then went slightly airborne as it exited. We traveled with the jeep topped out at around sixty mph for nearly two miles, until George began to relax a bit. But he was still on a high when he looked over at me and said, "I love Fords … I really do … do you realize I have an honest to gosh war story I can tell my grandkids; I really do!"

George slowed down after a few miles and we made it to Phu Cat in another fifteen minutes. Phu Cat was a US Base, so they had a small US Canteen and a PX. I collected the money for the pay at the US Pay Office, then looked up George and found him in the PX.

Just as I spied George, a shipmate of George's from boot camp walked up. "Yo, Georgie, old man, what's new?" said the young sailor. "Hey, what is this … old home week?" George answered. After some small talk I heard George ask him to join us for some lunch and a beer at the canteen. "No can do, old buddy, we gotta get back to Qui Nhon; I'm with my old Marine sergeant, he's got something going mid-afternoon, so we gotta scoot. Besides, we chowed on the road coming up. Ran into some Vietnamese sailors fishing with concussion grenades. Can you believe it? What a blast! Small pond and using grenades! Now that's my kind of fishing, old buddy! They invited us to join, so we served up the beer and they served up fresh fried fish. Man there was fish everywhere; I'm stuffed, old buddy." George became quiet and then asked tentatively, "Where were they fishing?" "Oh 'bout ten miles back down the road to Qui Nhon, right by a little bridge," he answered. I saw George's ears turn a little red.

George bid him goodbye, as they shook hands and parted. George and I avoided eye contact and said nothing. We walked to the canteen, no air left in our respective balloons. We drank some Pabst Blue Ribbon and ate some genuine simulated American hamburgers.

After a while George started to chuckle, and said, "Shoot, I can't believe it! I didn't see a pond. Did you see a pond, Lieutenant?" "Never saw a pond," I confessed. The beer began taking effect, and we both started chuckling, which quickly transitioned into near out of control falling down belly laughter. We sat there and had one last beer, and I said to George, "We need to scoot, old buddy!" which brought another round of laughter. After we settled down, I said, "George, what say we do a little fishing on the way back?"

"No way, GI … but I still love Fords … I really do … I'm still going to buy a Ford when I get back."

Quonset Life

When I was stationed on the Qui Nhon peninsula, I slept in a small quonset hut along with the other American advisors. Quonsets were all over Vietnam. The floor was a plywood structure that rested on cement blocks. The rest of the building looked like a big half-moon shaped hollow cylinder with the flat side resting on the plywood floor. There were sandbags piled up at both ends of this mostly sheet metal building. Inside the building there were eight little rooms separated by plywood walls that were about six feet high so that if you wanted to, you could jump up and look over them. There was a narrow hallway down the middle of the quonset and plywood walls on either side with narrow plywood doors and a rude metal latch on each. We called the quonset a "hooch." It was our hooch. We had a hooch-maid, a Vietnamese lady, we called her Mamasan, in our young insensitive manner of the day. She was probably no more than thirty years old, but she looked like a grandmother to us. She took our laundry in the morning and after hand-washing it on her hands and knees in the outdoor shower, returned it cleaned and neatly folded in the evening. All that for a few piasters of local currency a week, and we never gave it a second thought. One needed only two sets of clothes: one being worn and one with Mamasan. It was home-sweet-home and I was grateful to be there. It was certainly paradise compared to toting M16s in the Central Highlands.

Vietnam uses the Chinese calendar, having been influenced by

Old Killer Dog had a special affection for Tommy.

Chinese culture during long periods in the first millennium. For example, the Chinese New Year is celebrated as New Year in Vietnam as well, where the festival is referred to as "Tet." Calendar years run in cycles of twelve, with each represented by a Zodiac animal. When I was in Vietnam, it was the Year of the Rat. I was not necessarily inspired by this fact. Having grown up on a farm, I was well aware of the many non-admirable qualities of rats and what the evidence of a rat looked like in a corn crib or granary. I was sorry to say I had seen those familiar signs around our quonset. It would have been more pleasant, I felt, if our patron animal had been something like the Tiger or preferably the Dragon.

To further describe our Zodiac animal, "Rat" years ran in cycles; there were the earth, metal, water, wood, and fire years. In 1972, the Water Rat took its turn, which in my opinion only degraded the already unpleasant image. I preferred to think of it as a muskrat, from boyhood days, although I was told it meant something entirely different. Not that I'm complaining, but I must also mention that our own Gregorian calendar reminded me that 1972 was a leap year; Oh goodie. An extra day in Vietnam in the Year of the Rat. We were twice blessed!

LTJG Tommy O'Brian's room was across the hallway from me. Tommy was a naval line officer who was also a cop before he joined the Navy. He was at Qui Nhon to advise the boat captains on coastal and river warfare tactics. Many an evening we spent in his cubicle or mine commiserating over our lives as advisors, sipping spirits till the wee hours. We'd typically close down an evening concluding, "Is anyone doing any real advising around here and, by the way, who the hell would build a base here?" The base was on the water's edge of a thin peninsula open to attack front and rear: steep hills to our back with vegetation that walked up to greet us on three sides and water lapping up from the bay on the fourth side. The VC had their choice; they could come by land or by sea and wipe us out. It especially bothered me that the hooch was so close to the water. And twelve inches of air between our plywood floor and Vietnam soil seemed a perfect clearance for any would-be sapper to blow us all up. Sappers were VC saboteurs, and were revered in Vietnam by friend and foe alike for being able to penetrate concertina wire and base defenses, and their specialty, an ability to slither underwater along beaches and river banks carrying all kinds of troublemaking munitions and paraphernalia.

One night after I'd been there about a month, the base was hit by sappers. Our hooch was far enough from the attack so that no one was injured; but underwater sappers attached explosives to the swift boats moored at the main pier and destroyed them triggering deafening explosions and fires. We all grabbed our stuff, and dived into a bunker just outside the hooch. After that I slept in my clothes and boots (untied) with weapons and ammo close by. It was all for nothing, we never got hit again at Qui Nhon, but I slept better.

I was told that at the beginning of the war each cubicle started out with just a single bed. Subsequent occupants added and subtracted furnishings and amenities as they saw fit (and as they could scrounge), much of which was left behind when they'd completed their year-long stint. I learned that a past occupant of my cubicle was an architect; he'd left a drawing table and a stack of drawings. Among the stack I found a detailed drawing labeled "Hidy Hole" which described plans for an escape route to an underground bunker beneath my cubicle. I thought, "Excellent! I have my own back-up plan."

I checked it out one day by removing a steel plate on my floor, but found Hidy Hole half-filled with water. I was starting to get a little paranoid over all this talk about the Year of the Water Rat. A man could drown down there waiting for the attack to be over, that is, in the unlikely event that the attack wasn't coming from sappers below the hooch; in which case I'd be the first to be dispatched. I put the steel plate back over the hole. But then I started worrying about sappers coming up through Hidy-Hole during the night; Lord only knows where that water-filled passageway led under there. I took the steel plate down to the boat repair shop and drilled several three-eighth inch holes around the plate circumference, then screwed the plate back to the floor with lag screws. I felt a little more secure knowing that if someone was trying to come up through Hidy-Hole, they would make considerable racket trying to bust out those lag screws, and it would give me some warning.

Tommy's room gave one an interesting perspective to gauge just how long this war had been going on. Over the years, previous occupants of his cubicle had contributed to papering the walls with Playboy centerfolds. When I say papering, I mean all four walls … and the plywood door. Yep, every square inch; and in some places it was multiple layers. In fact it wasn't all that easy to find your way out of there if you'd been sipping spirits complaining about your life and the door was shut. All the walls looked the same. Without being too specific, the door was where Miss Somebody, 1961 was sitting on the hood of a '61 turquoise Ford Galaxy Sunliner. Had we really been there for eleven years? No wonder the world was weary of Vietnam by 1972.

In our hooch we had a faithful watch dog named "Killer." Killer was a mongrel of an animal if ever one existed; about the size of a small German Shepherd with an extremely ugly face, a mangy brown and black coat, and was probably not up-to-date on his shots. During the day he could be found asleep just inside the entrance to the quonset, or sitting directly outside the mess tent so he could take full advantage of hand-outs the generous GI's would toss him as they exited three times a day.

The result, of course, was a dog about twice the weight that he should have been, that waddled in a rather humorous manner and was really quite lazy, but very friendly to his doting GI's. I must say, however, that Killer would get nervous around the Vietnamese sailors just as if he understood their discussions regarding a canine specialty dish that was served in the finer restaurants in Saigon and Da Nang.

For some reason Killer had a special affection for Tommy, and after a certain night, we were all convinced that Killer was Tommy's special guardian.

It was a normal incredibly hot and sultry night. The lights were out in the quonset. The thick atmosphere hung over everything with no discernable air movement. Warm droplets formed and then dripped from the liberal vegetation that surrounded our abode. Sometimes in the still of the night one could hear the water drip (along with the familiar whine of pesky mosquitoes, near one's ear, of course). We'd learned the best way to fight the heat was to lie still and minimize the use of energy. As a cop before Vietnam, Tommy was pretty relaxed. He always slept only in his shorts, stretched out on his back, bare-chested; he'd turn a small battery operated fan on himself and drift off. It was so sweltering that he claimed it was the only way to go.

This particular muggy night, around 3:00 A.M., we were all snoozing, except the man on watch (we hoped). Killer was in his normal position by the entry door, dead to the world; the temperature had just started to cool down. The only sounds were the snoozing and snoring of the quonset advisors.

With no warning, there came in the darkness an ear-piercing

scream of such incredible magnitude that it made one bolt upright and then shudder from a huge shot of fear-induced adrenalin. About one second later massive barking commenced. Then came the lights, sounds of boots stumbling on plywood, ammo clips being rammed into M16s, doors slamming, and general ruckus. I bolted up ram-rod straight literally quaking in my untied boots. I thought, "Oh man, this is it! This is what I've been preparing for, sweltering in my clothes and boots every night so I'd be ready! It's time!"

I opened my door holding a loaded .45 pistol and realized the ruckus was in Tommy's room. His door was half open and I thought, "Son of a bitch! A Viet Cong got Tommy, probably a sapper with a knife, and now I've got to go in there and shoot the bastard! Tommy may still be alive!" I saw Killer was already in Tommy's room viciously barking, growling, tearing, and ripping in the far corner, and I thought to myself, "Good old Killer, he'd tried to save his master, and he'll probably take a knife in the gullet for it himself, what a great dog! Mental note to recommend a citation for Killer for trying to save Tommy's life when we send Tommy's body back."

It was still dark in Tommy's room and I couldn't see the enemy, I inched my way against the door, and then kicked it all the way open. By this time, some of the others came up loaded for bear and we all poured into the room and got the light on. What a mess in there! Tommy was alive! He was standing on his bed in a flak jacket and shorts, holding his M16 by the barrel and ramming the butt of the weapon into the butt of Miss September, 1964 as he scraped through several layers of wall paper. Killer was ripping and tearing at the same area. There were decimated playmates all over the room.

"What the hell is going on in here?" was everybody's question. "Rat! Rat the size of a raccoon! The sumbitch was sitting on my chest, staring right in my face! The bastard was so heavy … woke me up … couldn't breathe!" Tommy answered gasping for air. Someone standing near Tommy's door holding a loaded M16 aimed at the wall and just over the advisors' heads yelled, "Was it a water rat, Tommy?!" "No, just a regular dry rat," was Tommy's ridiculous but truthful answer.

The rat was fighting for its life burrowing itself under all the playmates trying to inch up the wall and get high enough so the dog couldn't reach him. Old Killer, in all his fatness, was taking runs at the wall trying to get higher in pursuit of the enemy. The wall was taking a great deal of abuse from Killer's full body slams and Tommy's ramming. Both combatants were out of breath. Tommy dropped the gun down on the bed, and let the dog work awhile. Somewhere the danged rat had hidden itself among all those playmates and was not moving. Things quieted down, the boys sauntered back to bed. The dog was reduced to growling, panting, and waddling around in the shredded Playmates while staring at the upper part of the wall where the rat was last seen.

After a while, with all quiet, Tommy believed the rat must have escaped and all the advisors were soon back snoozing. Killer wasn't so sure, although having just had his biggest workout in months, he was looking forward to rest. Tommy left his door open for Killer to have quick access should the enemy reappear in the room somewhere.

I would say the hour was about 5:00 A.M., just before dawn, when the rat must have realized his opportunity for escaping under the cover of darkness was disappearing. Killer may have lost some agility with all that fatness, but his sense of smell was Olympic, and when the enemy made his move to escape, Killer was on him in a flash, pinning him between the wall and the floor in the hallway; Tommy finished him off with the butt of his rifle without the distraction of the centerfolds to fault his aim. This time, although the sounds of the fight were brutal, everyone slept through it except the rat, the dog, and Tommy. Killer dragged the dead enemy outside the quonset and sat proudly with his quarry accepting the praise of all quonset inhabitants when we exited the hooch that morning. Someone grabbed their WESTPAC 35mm camera and got a shot of Killer and his prize. They pinned the snapshot up in the mess hooch with the label, "Year of the Rat."

About a week later, I was transferred to Saigon to work with the Vietnamese Navy Construction Bureau down there. I dropped in on Qui Nhon about a month after that, but Tommy had gone back stateside early, and they were starting to shut the base down. There were only three navy advisors left in the hooch. No one had bothered to sweep out Tommy's old room; ripped up dusty wall paper was still all over the floor. Old Killer was nowhere to be seen. I kind of missed quonset life, but not too much.

Flight on the Curtiss

It happened one morning in the Spring of '72. That's when Harlan and I ran into the American nurses transporting the Highland fighters. It was just a chance encounter as I took off from a field outside the city of Qui Nhon. It was the first time I flew Air America, the CIA operated airline in Southeast Asia. Sometimes they were called CAT, Commercial Air Transport. Navy advisors were allowed to tag along with Air America if there was space available, and I was trying to get down to Cam Ranh Bay.

Construction Mechanic First Class, Harlan Wolfe, our burly lead Seabee from Alabama was to drop me off on the west side of Qui Nhon City for my flight down to Cam Ranh. He was glad for the chance to get off base and make a morning of it in the city after I caught my flight. We got an early start so we'd be there when the aircraft landed. I'd been warned to be on time, Air America didn't hang around; they wanted you there when the aircraft stopped rolling. Their preference was to drop out of the sky, make the pick-up, and be gone in minutes before anyone knew they were there.

Harlan Wolfe knew the routine well. The aircraft would land at an abandoned soccer field adjacent to a grove of banyan trees which ran nearly the length of the field. Harlan had it all worked out. He'd circle around the City and come in from the west side of the grove on an old access road and wind right through the big old trees. He'd

When the pilot left the aircraft, the big bird stood empty.

pull the jeep up to the edge of the tree line, and stick his nose out just far enough so we had a good view of the field but were still under cover.

This particular morning we were very early. Harlan pulled the nose of the jeep up just right, then shut it down. He reached back and pulled up a box from behind his seat and sat it between us. He grinned as he pulled out a bag of cinnamon rolls, still a little warm, and a thermos of coffee he had requisitioned earlier that morning from the mess hooch. "Gotta keep up the reputation for Seabee ingenuity, Lieutenant, and besides, it takes a lot of chow to keep this gut in shape!" he chuckled as he massaged his girth.

We sat there, drank coffee and studied the soccer field. "You know, this is a nice flat field; it's a shame the kids can't use it with the war going on and all," Harlan mused, "but it does make a nice landing strip for the Curtiss." He explained the process of the landing. The plane was a two-engine "tail dragger," a Curtiss C-46 Commando that carried no markings. It usually landed from the left side of the field and would come bouncing by us, taking up most of the strip to slow down as its tail disappeared into a cloud of dust. The pilot would then spin the aircraft at the far end, causing the little rear wheel to scribe a quick half circle, then gun the engines back down the strip. He would shut the engines down approximately where he had first touched down. We were sitting on the "fifty-yard line," where the old access road ended, so we were to high-tail it along the edge of the field as soon as he passed us going left.

We sat there watching and listening for the plane as the time neared. Suddenly we noticed a gray box-type truck pull up next to a block building on the opposite side of the field. "Correct me if I'm wrong, Lieutenant, but aren't those round eyes in that cab … and I do believe they're members of the opposite sex. What are they doing up here?" The truck was parked facing us and wore a Red Cross emblem on the front above the cab.

Two young women of college age got out and walked around to the back. "Whoa! Americano round eyes, no less!" Harlan observed as he became more interested. Wearing American civilian clothes, we began wondering if they were nurses or perhaps aid workers. Except for the jungle boots they looked like college coeds walking across campus. "I may have to provide assistance to the ladies while you get aboard," Harlan smiled, as he took a big swig of coffee.

The discussion ended abruptly when Harlan hit the starter and cried out, "That's the aircraft! See, she's all silver!" He shouted over the engine noise. "Only ones you'll see around here that aren't some shade of olive," Harlan said as he hauled on the wheel. I agreed … no military markings, it looked every bit a civilian aircraft. Less than two minutes later the aircraft hit the field and bounced past us as Harlan had predicted. He started moving down the line of trees and then hit the gas heavy after the plane passed us at speed on the return taxi. He cut to the right at the end of the field and came up behind the aircraft.

We waited a minute or so with our faces turned downwind for the dust to clear. I grabbed my overnight bag and jumped out; Harlan joined me as we approached the aircraft. "Don't forget to pick me up tomorrow," I said, "Same time, same place." The pilot was already out of the aircraft and walking quickly toward the Red Cross vehicle when we came around the plane. "Be with you fellows in a minute," he shouted to us, but kept moving toward the truck. The first thing I noticed was that the pilot was in fact the entire crew of the aircraft. When he left the plane, the big bird stood empty. The second thing was that he had his full attention on something in that truck. As we rounded the aircraft, we came into direct view of the truck that had pulled up to within fifty feet of the plane with the back end of the truck now facing the side of the aircraft. Nothing could have prepared us for what we saw next.

There were four American twenty-something women; the two we hadn't seen had been in the back of the vehicle. As we approached, three of them were coming toward us carrying bundles of something heavily bandaged they'd just lifted from the tailgate of the truck. I couldn't make out what they were. At first I thought they may be injured animals or small children. As they passed by us, though, I realized they were carrying horribly mutilated men, Montagnard tribesmen ... terribly disfigured and barely recognizable ... most of their limbs were missing. They had obviously survived some horrific action and perhaps only now were stabilized enough to transport. The women were focused on getting the men from truck to aircraft as quickly as possible.

Clothed only by massive bandaging about what was left of their bodies, there were eight individuals; some were just torsos of their former selves. None had legs, only one had both arms; two had literally no limbs. All were alert and wide-eyed, but muted. I had never seen anything like this before. The extent of their situation was so grave that it stopped Harlan and I abruptly, and we stepped back as they were carried to the aircraft.

They carried the men as one would hold a toddler: one-on-one, chest-to-chest, one arm under the bottom, one around the back. The first pair passed by with the tribesman's lone arm clinging to the young nurse's shoulder to help support his weight, the next had two arms locked around his carrier's neck. The bright mid-morning sun and steamy temperatures accentuated the brutality they had suffered. The strength the women displayed ... mentally, physically, and emotionally was impressive as they carried the tribesmen.

We began to realize these men were the strong and fierce fighters from the remote mountain locales we'd heard about who had sided with the Americans. These were not Viets. These were the indigenous tribal forest dwellers. They were people who lived outdoors; rough, lean, scarred. Their muscular backs and shoulders glistening with sweat and their ruddy faces were a sharp contrast to the American women carrying them to the plane. The gruesome sight was made all the more surreal by their quiet obedience, and their total reliance on those they clung to. Harlan whispered, "This must be the war behind the war, Lieutenant." I nodded affirmative.

The Montagnards lived in the Central Highlands and other mountainous areas of Vietnam. They were fiercely independent from the main society and had sided with the Americans in the early days of the war mostly because of their desire for independence and the way the communist cadres had treated them. Because of their special knowledge of the land and forests they were valuable to the Americans for secret operations. This one had obviously gone horribly wrong for these men. Their massive injuries suggested land mines or booby traps were involved. We supposed their American contacts were the CIA and Special Forces, and that was why they were now being carried to a CIA aircraft.

The nurses went about their work. They brought the men into the rear of the aircraft and then walked up the inclined aisle to mid-plane, where they situated them and strapped them in. They surrounded them with blankets and made them as comfortable as possible. Three of the women stayed on board with the men; the fourth returned to the truck.

I waited until all were on board, then climbed in and nodded as the pilot explained, "We'll make a stop at Da Lat in the mountains first, then over to Cam Ranh." I sat in the tail of the aircraft directly behind the rear-most tribesman. Except for the man in front of me, I could see only the heads of the three nurses, and two other tribesmen that were sitting on aisle seats.

The pilot closed and secured the rear access door then took his place in the left operator's seat and started the engines. He gunned the Curtiss back down the soccer field and took off immediately. Looking down, I could see Harlan working his way back up to the access road then disappear in the banyan grove. We quickly got up to altitude and settled in for the twenty-five-minute flight down to Da Lat. The aircraft engines were noisy during climb-out but quieted down a little at altitude.

I sat and thought about what was going on here, and I tried to understand what had happened. The fighters' situation was about as desperate as any I'd seen. An incident like this was enough to remind anyone of how poorly we understood the complexity of the war over here. I strained to hear or see anything that may explain what had happened.

I wondered about the four women and if they were CIA employees. I should have asked, but it seemed like a stupid question at the time, so I kept my mouth shut. I doubted if many Americans back home would ever understand what the women in these types of jobs went through over here. Thinking about this, forty years later, I'm sure some of them must still carry scars of those days.

We'd been flying for ten minutes when a blanket from the fighter in front of me fell to the floor. I instinctively jumped forward in the aisle to retrieve it. As I picked it up I found myself face-to-face with the man. He stared directly at me. I could see he was a young man in his twenties. His dark face was lined beyond its years and held unhealed scars from shrapnel.

He was the only fighter with two arms, and he tried to ease his position by lifting himself against the armrests. His face was full of terrible anxiety. With his left hand he pointed to me; and tried to say something. I couldn't understand him, and shook my head.

"They don't speak English," the young nurse across the aisle spoke up, "French … they speak some French."

"I wish I knew French … what's he want?" I asked. "Sorry, too much noise," she replied, and it was clear he wasn't going to say any more. I didn't know what he asked and couldn't respond in French anyway, so I took the easy way out and gave him the old American thumbs up. I put his blanket back by his side. He nodded slightly. I sat back down. The man was in bad shape, no question, and I doubted that he had much longer to live. It occurred to me that the CIA may be transporting these men back to their original locales and homelands in the mountains around Da Lat to spend what days they had left with their families.

A few minutes later the aircraft landed at Da Lat. The nurses carried the wounded men off the aircraft and into a covered staging area. From the passenger window of the Curtiss, I watched as two unmarked ambulances drove up and prepared to receive the injured tribesmen.

The nurses repeated their careful business of moving the stoic men. The ambulances then departed with the mangled fighters and the American women. With that, my window into one of the ugliest but most compassionate sides of Vietnam closed … a glimpse of what human beings can do to each other and what they can do for each other.

Then the pilot wheeled the big silver Commando around and roared back down the short runway with just he and I aboard. We took off and circled out over the Pacific and headed toward Cam Ranh Bay.

The Man on Hill 131

I'd been watching him since he first started down the path on the hill in the distance. He caught my eye while I was scanning the hills surrounding Qui Nhon Support Base. It was 6:00 A.M., minutes after a spectacular sunrise that promised a beautiful day weather-wise along the Pacific coast of central Vietnam. I was working my way through the only drinkable coffee in the area, outside Trang's one-room café along the base perimeter wire, when I first saw him.

He was a solitary figure making his way slowly but steadily down the trail toward the north gate. At first it was just a movement, and I took it to be a dog or small animal meandering along the trail; but then I saw the light-colored shorts, like tennis shorts they wore in Vietnam in the cities sometimes. As the minutes passed I realized it was an elderly man shuffling along slowly, a bit stooped, but moving at a steady gait. I gazed through my field glasses; as he came into focus I could see shades of brown and gray that blended with the green vegetation. An olive shirt was tucked into his shorts that were drawn up high. He was small and his clothes could cause one to mistake him for a little boy from a distance. His face and bare arms were darkly sun tanned. His legs were bowed and nearly black so deeply tanned they were. Through the glasses I could see his rough bare feet kicking up little puffs of dust on the trail. I put the glasses down and continued to watch and nurse the coffee.

"There was fire on that hill last night, one ARVN killed," Xuan said.

There were no Americans on that hill any longer; they left in the fall of '71, and now a small detachment of Vietnamese ARVN soldiers stood guard over that highest point on the peninsula.

The peninsula was a strip of mountainous vegetation that blanketed a steep ridge of a backbone which ran the peninsula's length. The strip hooked to the southeast; and our little base was on the end of the hook. It flattened out to a beach at our end and made for a wonderful protected bay. The base was staffed by Vietnamese sailors, but we had a small contingent of US advisors there to provide guidance on maintaining the base and the operations out of the base. I was a Navy advisor there to instruct the Vietnamese officers on public works and perimeter defense. Hill 131 was just north of our base.

My Vietnamese counterpart was Lt. Xuan, in charge of the base facility. He was responsible for the generator equipment, boat repair, and perimeter defense, basically everything except the military boat operations that were initiated from our base. Xuan was capable but secretive, and I had first found it difficult to trust him. Some of the Americans said he was Viet Cong. I always thought he knew more about what was going on in the area than we did, than even our intelligence officer. He never crossed me, though, and, just two weeks before, he had warned me to stay away from the pier one evening which turned out to be the night someone decided to blow it up along with the swift boats. Sometimes Xuan would stay across the bay in the City during the night, but was always back early in the morning.

"There was fire on that hill last night, one ARVN killed," Xuan said as he sat down across from me. "Sappers turned the claymores around, a soldier was killed when they detonated the mines. The father of the dead soldier is coming down the hill now."

I thought to myself, again Xuan knows what happened before anyone else, and reasoned he must have been in the City last night.

I thought of questioning his sources about the man on the hill, but knew he'd only smile politely and be vague. He would never outright stonewall me. He would just not be forthcoming with details.

It was as if he was walking a fine line between us and them, and trying to manage the war all on his own. I felt he gave me only what he thought I needed to know, but I'd learned months before not to question him. Still, I detected a curtness and a preoccupation in him that morning and wondered if maybe something personally had gone wrong last night. Perhaps a family member or friend was involved in the incident on the hill.

I continued to focus on the man coming down the hill; he was squinting in the bright sunlight now, and I saw he was carrying something … a small object of some type. He was walking slower and I could see he was definitely heading toward the north gate of the base. Xuan said the man had walked the peninsula trail up the mountains during the night to identify his son. It was a much longer walk than coming through our base, but he didn't think the guards would let him pass through the base at night. "Maybe he thought his son was still a little bit alive and he could see him before he died completely," said Xuan. I thought to myself that was as good an analysis as I could have given.

The man approached the north gate and two heavily armed Vietnamese Navy guards met him. We watched them speak a bit, and watched through the glasses as the two inspected the object he was carrying; then, with one on each side of the old man, they escorted him through the gate. After a few minutes, the three of them started walking down the main dirt road on our base toward us. We put the glasses down. Xuan lit a cigarette and sat back in his chair. I finished my coffee.

I asked Xuan, "Why did they let him come on the base, it's not a good idea." He said not to worry because they know him, and added, "They probably felt sorry for him and let him pass so he could take the ferry back home to the city from our pier."

As he came closer, I saw his hair was a perfectly shaped gray flat-top right out of 1950s America. You'd think it would be in total contrast with his tanned lined face and deep set eyes, but somehow it fit and gave his face a distinguished appearance despite the ragged clothes. He looked straight ahead at Xuan and I, his face had no expression; I thought he may be in shock. The Vietnamese Navy men continued to bring him toward where Xuan and I were sitting. I could now see he was carrying a small paper bag that read *Dep* in big letters on the side, a discarded shopping bag from a lady's salon whose name means beauty.

When it became clear they were going to approach us, I asked Xuan what he supposed was on the man's mind. "That's easy: he hates the Communists, he hates the Americans. He's thinking his son never knew Vietnam without war," Xuan seemed to speak the words as his own sentiments as well as the old man's.

The three came directly up to us. We stood up, and I watched the old man closely as he looked up at Xuan and I. Xuan looked away. The man gazed at me but said nothing. He wore the resigned look of infinite patience and of a life spent in manual labor; wiry of stature, farmer strong, but his tired eyes were clear and gave the impression of wisdom; but not about to be shared with me. I said to him in Vietnamese that we were sorry for his loss. He bowed slightly and as he put his hands together in a kind of praying motion he asked to take the ferry. Then he cast his eyes down and stood there. I said to Xuan, "You should check that bag before you let him past the fuel station and onto the pier; with all those boats, we don't need more fireworks on the pier." Xuan said something in Vietnamese to the Navy guards. They answered back quickly without looking at the bag. Xuan turned to me, "The bag is no problem."

The Navy men asked again about him taking the ferry to the city. Xuan turned away and stared at the water. The old man turned and looked at me. I nodded to the men to go ahead. The old man bowed slightly.

We watched as the Navy men helped the old man into a PBR patrol boat and tossed the lines back onto the pier. The old man sat in the bow of the boat and stared back at Hill 131 as the boat sped away for the ten minute ride to the city pier.

I asked Xuan what was in the bag. He turned to me and said, "Nothing of importance, only some old photos of the man's son and his son's wife." Then Xuan turned and stared back at the water. His eyes had glazed over. After a while, he said to me almost under his breath, "My father has lost another son."

High Seas Adventure

A young boy in Ho Chi Minh sandals scrambled up a step ladder and changed the flight schedule. I watched and winced ... the aircraft to An Thoi had been cancelled ... for the third day in a row now. The reason given was "Eastertide." Aircraft were needed up North for what they were calling the "Eastertide Offensive," the action begun by North Vietnamese regulars bursting over the DMZ and taking the city of Quang Tri.

I was at the Can Tho airport waiting for my next flight on a steamy morning after hitching a chopper ride from Saigon. Can Tho sat firmly in the middle of the Mekong Delta and was a stop along my way to An Thoi Navy Base on Phu Quoc Island. I was heading down there to help the Vietnamese Base Commander with "Vietnamization," a word recently entered into the lexicon of the Vietnam dialog ... the "secret program" to end the war that Nixon unveiled after his landslide election: turn everything over to the Vietnamese, including the Navy bases and then bolt and declare victory (those weren't his exact words). Phu Quoc Island was part of South Vietnam and therefore its Navy base at An Thoi was on the list. It was just under a hundred miles out in the warm waters of the Gulf of Thailand off the west coast of South Vietnam.

With no aircraft available, it was back to Saigon for me and try again tomorrow ... or was it? Being the non-hero that I was, I was

Within the span of thirty minutes, there was no land visible in any direction.

loathed to continue flying back and forth across the Delta if unnecessary. Those dang SAM missiles made my blood run cold. I needed another way to get to An Thoi. I bought a bottle of room temperature Sprite from a portly mamasan with a smiling mouthful of shiny onyx teeth and sat down to cogitate.

The sultry morning was beginning to get really uncomfortable, but my spirits brightened a little as I glanced down the flight line and saw a man in green fatigues and a red beard coming toward me.

"Well looka here!" he said, "What in the world are you doin' here, Lieutenant?" Sure enough, it was Lyle Fayette all the way from Jackson, Mississippi. Lyle and I had done our advisor training together at Mare Island back in California and had become good friends. I explained my need to get to An Thoi.

Lyle was a wiry, savvy, and good natured old sea dog. A Petty Officer, Third Class and Navy veteran of twenty plus years, he'd been around the Horn for sure and thrived on an old school life of whisky, hard living, and making a profession of avoiding work. Lyle could be a hot-head if provoked, though, and didn't have much patience for authority, hence his third class rank after twenty years in the Navy. He'd survived ice storms in the North Atlantic and fought his way out of bars in the fleshpots of Asia and kept coming back for more; he was coasting through his year in Vietnam.

Lyle had the most rubbery face of any man I've ever known. A man of a thousand expressions, his skin didn't like the sun so it was always red in Vietnam; but it went well with that thick rusty beard. Overall the pliable face lent just a hint of Popeye to his countenance. Lyle had no fear of anything that I know of. He enjoyed and knew a lot of jokes and loved to laugh. But that face was something. He could look as sorrowful as Mother Theresa, or as crazy as Jack Nicholson within the measure of a second. He mostly wore his "like I really care" face. You could read Lyle's mind real time just by watching his facial expressions. I was a young man then, and considered Lyle as somewhat "elderly" at maybe forty, maybe forty-five?

It was good to see Lyle, but even better when I found out he was a swift boat advisor at An Thoi and was trying to get down there just like me. Always the scrounger, he had found a way, although it didn't sound all that solid to me. "You wouldn't be up for a little adventure now would you, Lieutenant?" he baited. He sat down with a glass of tea, took a long slug then explained what he'd come up with.

Lyle had run into two young VNN sailors he knew from An Thoi at a rather unsavory downtown establishment in Can Tho. They were also on their way back to base and planned to take a commercial bus overland to the coastal city of Rach Giá. From there they would take what they referred to as a "lumber boat" across the Gulf of Thailand to An Thoi. Lyle explained, "Not sure what they mean by lumber boat, but I'm guessing it's a freighter packing building supplies into the island; said they've done it before. I reckon it's OK; after all, them boys are native, they know the lay of the land … I personally don't know squat about this bus business, but I'd trust them more than flying back and forth across the Delta." We must have been bored or maybe just plain stupid and decided to join them.

The bus resembled a short-coupled 1950s rural school bus. It appeared to be blue in color under the quarter-inch layer of good red delta bottom land and was loaded with lots of rural folks and the sweet smells of Asia. There was a cargo rack on top bearing bags of grain and some cages of chickens.

Lyle and I climbed on board and dropped our tickets into a basket hanging from the door near the elderly driver. He was a friendly old gentleman and bowed slightly to us as we got on board. We sat on the left side of the bus. Lyle next to the window, me on the aisle. Lyle clicked the school bus window down as far as it would go to get some air as the travelers filed on. The big red beard slowed traffic down the aisle a bit, but eventually all were crammed in and we began to creep out of the dusty station and make our way out to Route 17, grinding a pound of transmission gears as we proceeded.

It wasn't long before we were in lush countryside crossing small bridges over canals. Occasionally we'd get a good view right down a canal as it extended across the land, straight as an arrow disappearing into infinity in a mist of humidity. The Delta was flat as a pancake, and except for the road there seemed to be water everywhere: in the canals, rivers, and paddies. We gazed across the passing landscape, and watched the ancient agrarian economy at work in calf-deep flooded paddies.

As we pushed our way along the road to Long Xuyen, we passed a variety of small vehicles and people walking alongside the road. Lambretta scooters seemed to be the preferred two-wheeled vehicle often loaded with a family of four or even more. All types of goods were also being hauled on those scooters and motor bikes: building materials, fresh produce, even hogs and chickens were somehow transported heading to or coming from market. Three-wheeled Lambretta taxis carried passengers in covered luxury with bench seats facing one another.

Several young entrepreneurs had set up roadside stands with two stacks of bottles in little pyramids, one of Coke and Orange Drink and one of gasoline and motor oil in pop bottles for the two-cycle scooters. Often the gas in the bottles was displayed in bright colors to look more attractive and to lend credence to the seller's claim of extra power. I've often wondered if anyone ever got the drinks and gas switched.

The old bus creaked and groaned on, over ruts and little bridges and through the heat and humidity, until about two miles past Song Ha where the engine began to sound distressed and then died about one hundred yards later, having run out of gas. Most of the travelers exited the bus and we sat down under some shade trees. The driver didn't seem to be too concerned about the predicament as he pulled out his hammock for a snooze. Lyle and I were discussing the possibility of hitchhiking when two boys who looked no older than sixteen came along in a three-wheeled Lambretta toting a tank that looked to hold about ten gallons.

After some discussion with the bus driver, cash exchanged hands and one of the boys grabbed a hydraulic hose from the front seat of the scooter. He stuck it in the ten-gallon tank, put the other end in his mouth and started sucking, all the time walking toward the bus' gas cap. When gas reached his mouth he spit it out quickly and rammed the siphoned gas hose into the bus' gas snout. Lyle remarked with a deadpan face, "Modern science."

At every small hamlet, kids would surround the bus and try to sell us drinks and local snacks through the bus windows. They would gather around Lyle's window when they saw the red beard and realized we were Americans. *"Ong Mi! Ong Mi!"* they shouted and laughed, "Americans! Americans!" Lyle returned the favor with a hilarious assortment of expressions, which delighted the young crowd. He could even do an impersonation of Ho Chi Minh, which, I must say, was really quite weird.

We stayed on Route 17 heading north along the southerly flowing An Giang River, which forked off the Mekong and continued into the South China Sea. When we got to the town of Long Xuyen, we turned left and took Route 8, a straight shot west to Rach Gia.

Route 8 was long and bumpy and the heat was stifling. There

was low-hanging vegetation along the road which helped a little; occasionally we'd pass a car or motorbike going no more than fifteen miles per hour. Once, a convoy of ARVN military vehicles passed us at high speed which left us in a cloud of dust for several minutes. After that, we creeped along with most of the bus asleep until finally pulling into Rach Giá. To our chagrin, we realized the better part of the day was gone.

Rach Giá was hot and muddy after an afternoon shower. Within minutes we realized the Vietnamese sailors had taken off and were nowhere to be seen. The heat of the afternoon was upon us and steam rose from the street. We were uncomfortable and starved, and therefore overjoyed to find relief in a small mom and pop restaurant where we attacked two heaping mounds of fried rice and shrimp along with 33 Export Beer.

Rach Giá was nestled in a half-moon bay that opened to the Gulf of Thailand on the west coast of Vietnam, perhaps fifty miles south of the Cambodian border. It wasn't much of a town back then and it was easy to spot the piers and small harbor, so we headed there in search of the "lumber boat." Soon we located her and were disappointed to discover the "lumber boat" was in reality an ancient wooden ferry boat. It was loaded to the gills with local and regional passengers traveling to the Island. A decent sized vessel, there were three decks with open sides, perhaps built to carry two-hundred people. The boat appeared to be setting low in the water and passengers were still boarding.

Lyle and I purchased passage and hurried on board the crowded vessel. Several entire families were sprawled out on the decks, wherever they could find space. Those who could obtained passenger seats, many retired onto their bags and luggage. The scene made me visualize headlines such as, FIVE-HUNDRED PEOPLE LOST AS FERRY CAPSIZES IN GULF OF THAILAND.

There was no room on the two lower decks so Lyle and I climbed to the third. After two beers, and two flights of stairs, Lyle was huffing and puffing as we reached the rarified air of top deck. He was crest fallen and caught the eyes of two matronly Vietnamese women cooking supper for a brood of little ones and they beckoned him over to sit and have some tea. Lyle returned the invite with a huge pliable smile and sat down between the two of them. He looked straight up at the sky and said in his friendliest, "Let's hope it don't rain!" Everyone laughed as though they were old friends, having no idea what he said.

We settled in for the trip but soon learned that the ferry would not embark until 8:00 P.M. and was a slow overnight trawler of a boat that wasn't due to arrive in An Thoi until 7:00 A.M. It seems we had missed the earlier ferry the sailors had planned for. About this time Lyle spied the sailors on the opposite end of the top deck. We could see they were leaning over the top rail and talking with others on the grassy space between our pier and a small adjacent canal. Soon the sailors came rushing toward us as they made their way to the lower decks. They stopped long enough to explain to Lyle a change of plans: a Vietnamese Navy ship was about to cast off for An Thoi and they had just secured space on board. They recommended we follow.

The vessel was a coastal patrol boat that was referred to as a ferro-cement junk. Lyle's face began searching for the correct expression of fear until it finally telegraphed, "Cement sinks, don't it?"

The name sounds like something from a junior high science project, but I was somewhat familiar with the navy junks as I had seen some being built at the Saigon shipyard. The hull really was made of concrete … a molded thick matrix of concrete, chicken wire preform, and light rebar. There were two engines: six-cylinder

Detroit Diesels side-by-side each driving an industrial water pump that sucked in sea water, pressurized it, then blew it out at the stern under water. The two big pressurized streams of water were used for steering as well as propulsion.

The junks were about sixty-five feet in length and were fitted with a pilot house and plenty of fire power. They were low-tech, but they were heavy and strong and very sea worthy. The Vietnamese had mastered the art of manufacturing them quickly and inexpensively to render a reliable boat. When they plowed through the water with all that weight and ammo and those Jimmy's growling, they meant business.

When I heard it was a Navy junk, I told Lyle this is a no-brainer, let's jump ship. We immediately exchanged the lumber boat for a cement one. The young sailors explained who we were to the crew of four men, three young sailors, and a rather older and mysterious captain: a fellow named Loan, who offered us less than a gracious welcome. He seemed a lone wolf of a man, preoccupied and aloof from his crew. In fact, he wasn't wearing a regulation uniform and Lyle glanced at me with an old sea dog's wariness and a stern frown which said, "There's some risk here, but I think we could handle this little bastard if there's trouble."

By the time the junk got underway, the sun was westering. As we motored down a beautiful little canal that ran through a park on the west side of Rach Giá, the palm tree shadows were lengthening and pointing opposite to the direction of our travel. They resembled giant arrows that seemed to say, "You're going the wrong way." Lyle joked about it, but privately we both thought it an ominous sign, and doubts about this trip began to creep in. The canal had well-built vertical stone sides and was perhaps twice to three times the beam of the vessel so there was adequate clearance for the boat. But it still felt as though you could jump from the deck of the junk to dry land in an emergency. We joked about our last chance to terminate this adventure with one giant leap for mankind, but we stayed on deck and watched the park pass into our wake.

Continuing west we entered into the great embarkment of land to sea where the relative safety of greens and browns in forest and field succumbs to the unknown expanse of a blend of blue, gray, and white. Here, the indigo hues of the sky meet and mix with the copper green mood of the sea and the exact line of the horizon becomes lost. It was not a comfortable feeling heading out as we realized our options, should trouble arise, were rapidly diminishing.

Phu Quoc Island was due west from Rach Giá; at the predicted sixteen knots we were looking at five hours, which meant we'd be feeling our way into the channel at An Thoi as darkness closed in. I didn't like the idea of coming in there after dark, but didn't think it would be a big problem and maybe we'd get lucky and make it as dusk lingered. I watched the crew study the map and set a course which made us feel a little better that they knew what they were doing. But Lyle and I noticed the captain was not involved in determining the course and continued to be remote from his crew of three, never turning his back on them with his furtive eyes ever in motion. We decided to keep a watchful eyeball of our own on him.

The Jimmy 6-71s hummed away and the heavy junk rode steady and smooth as we made our way steaming directly into the sun. Within the span of thirty minutes there was no land visible in any direction. A sky of blue with only high and scattered clouds, a gleaming sea of deep blue-green with a hint of white caps in the extremes of vision and white gulls leading our way at perhaps one-hundred yards off the bow; this was now our world as we continued to plow smoothly and steadily. Lyle and I lounged on the starboard bow and finished a jar of

Planters Roasted Peanuts that Lyle had produced from his cargo pockets. A few more miles and the gulls turned back sensing approaching evening. Then we were completely on our own.

We motored steadily along at constant speed for three hours, the only change being the sun continuing to wester and the wake becoming more defined as crisp white compared to the sea as the sunlight moved from white hot to a yellow glow and the breeze cooled down.

Lyle heard it first: a cough on the starboard diesel. Then smooth running for another minute, then another cough. The port engine then sputtered and within seconds both engines were sputtering and losing power. The sailors quickly throttled back the engines, and the boat pitched forward. They disengaged the diesels and tried to rev the engines but to no avail. They then shut down the engines and we were suddenly adrift. Lyle looked at me and growled, "This ain't good."

Lyle and I went aft and watched the crew investigate the problem. After they had a brief look at the equipment, it seemed they all agreed on the cause of our misfortune and the rest of the discussion became an animated argument of shouting and cussing. A few minutes later the crew communicated to us quite in detail what the problem was in spite of the fact that none of the crew could speak more than a couple of words of English, and Lyle and I spoke only the basic of basic Vietnamese.

* * * *

To explain this I need to share with the reader a language code that was often used by GIs and Vietnamese in those days. The first part was the words: number one, number ten. Number one meant good, number ten meant bad; one was smart, ten was stupid; one was beautiful, ten was ugly; one was strong, ten was weak; one meant certainly, ten meant under no conditions. You get the idea. These elegant words were often supported and emphasized by the equally elegant descriptors: beaucoup and tee-tee. These were words of quantity and measurement: beaucoup of course being large, great, many, tall, thick, etc.; tee-tee being little, small, a few, short, thin, etc. The four words were often used together as in: *Beaucoup number one problem, tee-tee diesel!*

* * * *

Lyle and I stood on deck as two of the crew came up to us and launched into a five-minute explanation of the cause of our dilemma. Using only the four-word code and hand signals on both sides, we found out that we had run out of fuel, that the fuel from the boat had been siphoned and sold in Rach Giá by the crew, that they got a good price for the fuel, that the money was spent on an excellent meal and that after the meal they bought a case of beer and became very drunk and then were entertained by two excellent ladies. Overall they thought they got good value for the fuel, and hoped we didn't think too badly of them.

Lyle looked at me and said, "Well damn me, Lieutenant, I'm just so happy for them!" With an accentuated disgusted look, he went forward and lay down on the bow deck, reasoning that the most productive thing he could do at that point was to conserve energy. The captain was strangely uninvolved with these discussions and seemed only to want to know the conclusion.

The sailors told me they were trying to raise their buddies on the radio to bring a re-fueler. They estimated the fuel would probably be there sometime during the night. The captain then came over and asked how much ammo we had, since pirates often patrolled this gulf and sometimes stopped ferry boats or private vessels. He doubted that they would bother us due to the fire power on board, but we should be prepared. We were wary of his question and were vague in replying. Actually, Lyle and I both had only .45 caliber pistols. I had two clips, ten bullets. Lyle said he had seven, not counting the one he may use on the captain. As for the junk, there were two imposing M60 machine guns mounted on deck amidships at the ready with bandoleers hanging menacingly.

We noticed none of the sailors carried guns that we could see except the captain who wore a pistol in a little holster on his right side … not a military issue, a fancy civilian type. It looked like that one in the famous photo of the mayor of Saigon summarily and publicly executing a captured VC. There were also two large wooden cases in the control house. We assumed they were loaded with an assortment of automatic weapons, but we noticed the boxes were locked. From what we could determine, only the captain and Lyle and I carried weapons.

Darkness closed in around 10:00 P.M. as we bobbed on the Gulf of Thailand. The sky was dark dark, and the moon had a golden ring that produced an eerie light against the cement deck the color of marble, and the silvery sea was a carpet of fluorescent white caps. Lyle's rubbery face adopted a ghostly ash in the moonlight as he looked at me and croaked, "This here's downright spooky!"

At 11:00 P.M., the sailors and crew cooked some rice laced with *nouc mam*, a fermented fish sauce, the odor of which made Lyle sick to his stomach. He looked at me with a face of an apologetic puppy and said, "This here trip's a mistake and I'm sorry as hell about it, it's my fault." He looked real sorry too, and I told him he was a wimp which brought a little smile. He walked around a bit and felt better.

We shared a package of stale crackers and peanut butter I found in my bag from a previous trip and a can of some unknown drink the sailors gave us. I cursed myself for not being more prepared for this trip. Then Lyle went over and sat down on some gear on the bow deck and started singing this song:

Well it's red meat when I'm hungry,
And whisky when I'm dry,
Greenbacks when I'm hard-up,
And religion when I die ...

Two sailors with a small lantern were curious and gathered around him on the deck as he sang in a hoarse melancholy voice and stared out across the dark sea. An American salt singing a sea chantey. The young sailors seemed mesmerized by Lyle's lonely song ...

The whole world's a bottle,
And life's but a dram,
When a bottle gets empty,
Well it ain't worth one damn ...

By midnight most had drifted off to sleep, sprawled about the decks, except for the two beside the small lantern in the pilot room minding a crackling radio. I couldn't tell if the captain was asleep or not.

I don't remember what I dreamed about that night but I do remember the rude awakening. Around 4:30 A.M. Lyle and I were awakened by both M60 machine guns going off at once. One was only a few feet away from where we were sleeping. We were on our feet in seconds crouched over the stupendous fire power of our .45

pistols. When we came to our senses we realized two of the crew were firing tracer bullets almost straight up in the air to give our position to the re-fueler that was now in close proximity. Hundreds of streaks of red were shooting up into the dark sky and disappearing into infinity.

There was a lot of lead going up there and I was strangely taken back to my high school English teacher, an ancient Miss Marybelle McConnel, standing in front of our class reciting the words:

I shot an arrow into the air,
It fell to Earth
I know not where …

That lead had to go somewhere, I thought. It would eventually fall back down to Earth also, and I was hoping it didn't fall on me or Lyle … or the re-fueler.

I looked at Lyle, but he wasn't thinking about his English teacher. His face had turned dark and contorted and mean. He bellowed to one of the gunners, "That there's a waste of ammo and tax-payer's money, don't you imbeciles have flares?! Shut them guns down … NOW!" Now I personally wasn't inclined to shout negative orders to someone behind an M60 wherein only a matter of a few degrees of angle could mean the end of life as we know it, but like I said, Lyle was fearless.

The gunner stopped firing; I don't think he knew what Lyle said but Lyle's face and puffed up body language was frightening. The gunner walked away toward the fuel bilge with his tail between his legs and helped the others in tying the tanker up alongside to begin fueling. In a few minutes after the fueling started the captain walked up to the deck slowly and over to where we were; he looked at us askance with a hand on his fancy little holster. Lyle stood up and glowered at him with threatening dark eyes. Lyle looked as though he was ready to throw back a can of spinach and attack, but the captain said nothing and sauntered on back to the pilot house.

The crew had to disconnect the fuel lines and drain the air from the filters since they had run the engines completely out of fuel. Air in the line of the port engine managed to delay us another thirty minutes after re-fueling. As we finally resumed our journey, early dawn was beginning to break through in our wake and things started looking a little brighter.

Lyle and I were beginning to get real hungry again as we plowed on. The morning sun brought the heat and we stood on the bow deck in the breeze observing as we tried to stay attuned to the captain, the vessel and hopefully the approaching land. We began to notice strange behavior between the captain and the crew again. He had sent the two sailors from An Thoi over with us and he was having harsh words with the crew. We assumed he was upset over the fuel problem; the sailors from An Thoi were not involved in what was going on and were just anxious to get back to base.

The morning passed slowly. Around noon we turned our attention to scouring the horizon in earnest, and saw some sea gulls; then a few minutes later Lyle spotted land. He recognized it as Phu Quoc Island from the configuration of the mountains and the size of the land mass. It was now afternoon, and more than twenty-four hours since we'd left Can Tho. The only thing that stood between us and solid food now was about five miles of dark blue water. We plowed another thirty minutes before the crew cut the speed as we approached a small inlet by a coastal village. We noticed the captain had taken over the helm but kept the crew back with him.

"I'll tell you one thing, this here's the first time I seen him do any captain'n'. But there's something fishy about this scoundrel," Lyle postulated. I'd never been to the base at An Thoi, so I didn't know what to expect, but as we entered the small inlet, Lyle's face began to tighten up and he remarked, "This here captain must be coming in the backdoor. I didn't even know these piers existed."

We pulled up near a small dilapidated pier barely above water level and a good distance from the normal mooring area for boats this size. One of the crew indicated to Lyle and I that we had to get off here. The pier looked old and abandoned. Lyle looked around and said he recognized the maintenance buildings for the Navy base in the distance so we decided we'd better get off. The crew wasn't talking, but they positioned a long plank from the vessel to the pier and we balanced our way across and down the bobbing gangway along with the two An Thoi sailors.

As we walked away we looked back at the junk and noticed the crew of three sailors was still onboard but they were lugging one of the locked wooden crates up to the make-shift ramp to hoist it ashore. It seemed odd that the captain had gone ashore and was walking up the road into the village, but we thought it best to keep moving, and headed toward the Navy base.

Lyle and I trudged down a beaten pathway that ran along the shoreline perhaps a quarter of a mile to the main gate of the base. My body was still rocking with the ship's movement as we checked in, and then went straight to the advisor's mess hooch. It was now mid-afternoon and we were beyond starved. We dropped our gear and an American advisor Lyle called Cookie Monster rustled us up some navy powdered eggs and fried some ham slices he had squandered from the Fleet. He threw in some cold pineapple the locals provided and poured us our fill of cold iced tea. I remember it as one of the best meals I had in Vietnam.

We lingered there for an hour or so until we got our land legs back; then I headed to the Senior Advisor to see when I could meet with the Base Commanding Officer. I walked along with Lyle to the swift boat operations shop on the way since he wanted to tell the Vietnamese Operations Chief about the cement junk. Lyle's mood had changed to being more relaxed now that he was on his "home" turf, and he chuckled about the night on the Gulf of Thailand, but his face turned stern when we spoke of the ship's captain.

We found the Vietnamese Operations Chief under an elevated swift boat checking the propulsion ports. Chief Phan was a good natured gregarious sort and shook Lyle's hand vigorously when he saw him, "The sailor is home from the sea!" and laughed heartily. "And your first experience on a ferro-cement junk!" The two men obviously had a good rapport. "You bet I'm back; gettin' too old for this kind of travel, though! I expect the boat along with the crew will be tying up in the normal spaces shortly. Captain Loan appears to have gone ashore up by that little village."

Chief Phan's face became stern, "Mr. Lyle, that was not one of our boats. While you were eating, the crew brought that boat over here at our pier for a short while. They were a crew of three young men and were quite distraught as they explained their presence. They reported that they are stationed in Rach Giá and operate along the west coast of the Ca Mau Peninsula. I verified this with our Commanding Officer, Commander Tien. They also explained who Mr. Loan was, and why they came to Phu Quoc.

"They have left already, Mr. Lyle; they were in a hurry. Commander Tien authorized refueling them and providing provisions. But the man Loan you speak of was no captain. No sir, Mr. Lyle, he had his own plans. The crew told me Loan took their weapons and threatened them before you Americans came on board, but later paid them handsomely for the unscheduled trip to Phu Quoc." Lyle's face became red and he blurted, "I knew that scoundrel was up to something!" But Phan cut him off, "Patience, please, Mr. Lyle," as he smiled and kindly prodded his friend, "You'll never become an Asian with that temper!"

Phan looked straight at Lyle and continued, "You see, Mr. Lyle, you carried a cargo with you last night ... a locked wooden crate on that boat bore the young brother of Loan. You brought a young ARVN soldier that fell at Quang Tri home last night; home to Phu Quoc for the last time."

The Houses at Cuu Long

When I was reassigned to Saigon from Qui Nhon, I was introduced to two new counterparts: Nguyen Dinh Chu and Phan Dinh Tran. Vietnamese names typically have three parts: the first name is the family name. The last and sometimes the middle names are given names. So I addressed my counterparts as Chu and Tran. These men were both naval officers and both worked out of the VNN Construction Office located at the Navy Shipyard in Saigon. Typically Chu would handle the construction and other engineering projects in the Saigon area, while Tran accompanied me when the projects required travel across the country.

Of the two of them, I was closest to Chu and he turned out to be a trusted friend in Vietnam. He was a civil engineer, trained at Saigon University. Chu's wife was a primary school teacher in the city. Chu and I often traveled together, me on the rear of his Honda CB100 motorcycle. He was an officer the equivalent of LTJG in the US ranking system. We were alike in many ways: both engineers, both recently married with no kids, both wives were school teachers, both doing what we thought we were supposed to be doing. We also had a similar temperament, and enjoyed working together. Chu and his wife introduced me to restaurants and the colloquial language and culture of Vietnam. I spent time at their home and was invited to family gatherings.

The young girl hoisted the cement onto her shoulder and carried it, barefoot of course.

We built a lot of houses together, Chu, Tran, and I, during the year I was in Vietnam. The houses were a part of the project called the Dependent Shelter Program for VN Navy families. It was part of Nixon's plan of "Vietnamization," in particular the part to strengthen morale of the VNN by providing housing for VNN families that would allow them to live with the service member near the Navy base. We normally built the houses in rows of ten units, depending on the landscape. As row houses, they were more like condominiums than separate houses. We located them in little communities that were along the rivers and seacoasts. Overall, the program included nearly two-thousand individual units. These dwellings, with a main room, a loft, an open kitchen, and bathroom were bare-boned, and most Americans would hardly consider them houses. But to young families of VNN service men, they were comfortable and much sought after. We had long waiting lists for occupants. I was always amazed at how appreciative they were with so little.

On a bright morning in the fall of 1972 I found myself driving up the road from Cat Lai with Chu. Cat Lai was just south of the city of Saigon. We were working on a housing project on Cuu Long Island (considered a suburb of Saigon at that time across the Saigon River) and we'd made arrangements for building materials from some of the local contractors. It was one of the last projects that Chu and I worked on. We were following a supply truck headed for our project site; the truck belonged to a small company owned by a Chinese business woman who was supplying cement and tile for the project.

The two-ton flatbed ground along the two-lane road slowly in front of us, following the river. It carried one and a half tons of cement mix in heavy dusty bags. Standing straight and tall on the rear of the truck and steadying herself on the stack of cement bags stood Ming Cai, a youngish Chinese business woman and entrepreneur of about thirty-five years of age. She wore a white hard hat and dark sunglasses as she surveyed the road ahead and searched for her workers' pick-up point. "Get a move on, Xao!" she shouted to her driver, "The morning is half gone, and no work crew yet!" Xao gave her a frown and mumbled something she couldn't hear. Nevertheless he advanced the throttle of the ancient Mercedes and soon it was bouncing along at 35 mph which started kicking up some fine mortar mix dust. Ming pulled a scarf up to her glasses to shield her face. Chu and I dropped back a bit to avoid the small dust cloud they were generating.

They rolled past a shanty town of cardboard and corrugated metal shacks pressed up against the roadway, and then crossed a bridge over a slowly flowing small tributary, murky gray in color, that wound through the shanty town. The stream served as a source of water for washing and cooking upstream and a means to carry away waste on the downstream side. The short bridge trembled under the weight of the truck, but Ming was focused on the road straight ahead and ignored the bridge and the collection of humanity as they passed. At the next crossroad she found what she was searching for. She pulled her scarf down and shouted again to Xao, "Pull over, that's our crew!" The old truck ground to a stop a hundred feet or so past a little group of four people. They trotted to catch up to the truck and jumped on. Each found a place to sit on the side and rear edges of the truck's flat bed and the vehicle continued on its way to the Saigon suburb of Cuu Long Island.

Among the group were two young men perhaps in their early twenties, very rough looking, obviously day laborers. They were barefoot and wore old work clothes. Though young, they had the look of broken men weighted down by long years of hard labor. An older woman needed to be helped by one of the men onto the truck. She carried a basket that I thought could be their lunches and water for the day. She certainly didn't appear able to do much cement work. She was thin and sat bent over resting on the basket upon her lap, her face covered by her conical hat. The fourth person was a young girl about sixteen or seventeen years old. She was barefoot like the men, but her black sateen trousers and long-sleeved white blouse were clean. She was tall and thin and she stepped lightly on the ground as she approached the truck and sprang effortlessly to her place on the flatbed. She carried a long white scarf in her hand which she held up to her face as the truck began to move out. I thought to myself, "What a life; dirt poor, living in a cardboard shanty town, and scratching out a living from hit-and-miss day labor jobs."

We drove on, following the truck past industrial areas along the seaport and the commercial docks as we entered the city of Saigon. Chu pointed out the great granaries along the port that were commandeered by the occupying Japanese during WWII, and were used to store the rice harvest from the fertile Mekong Delta. Vietnamese farmers and landowners alike were forced to surrender their harvest to supply the Japanese fleet with food, while the people of the Delta starved. Chu said his father, a young man then, told of thousands that starved to death in Saigon when most of the rice was diverted to the massive Japanese Navy by way of Saigon through sea lanes out the South China Sea. "Not a leaf was to be found on any tree in the city," was how Chu said his father put it. All had been stripped clean by the population in a desperate quest for food. Many in Saigon still carried an animosity toward Japan and the Japanese people from that time.

We passed a small park with an impressive monument to Tran Hung Dao, the patron saint of the Vietnamese Navy, and a figure of hero worship throughout Vietnam. The statue presented him in typical pose, one hand on his scabbard and armored girdle, and the other pointing the way forward. There were many heroes in

Vietnam legend and history, but he was the greatest. In the thirteenth century he repelled three major Mongol invasions with victories over the forces of Kublai Khan, once destroying six-hundred ships of the Mongols' navy. We stopped and Chu snapped my picture in front of the statue. On we drove and quickly caught up with the cement truck. Soon we approached the bridge that crossed over to Cuu Long Island.

Island is a bit of misnomer, as Cuu Long was a separate land mass only by virtue of the kink in the river as it snaked severely and along with other channels in the river delta it captured Cuu Long as a land mass within the snake. We crossed a bridge and entered into the crowded suburb with its maze of small streets, most of hard packed dirt and stone. We continued on at slow speed as the old truck worked its way toward the small Navy support base further along the river where it began to widen noticeably. We followed and I watched as Ming Cao spoke impatiently to her driver as she stood over the passengers. The old woman appeared to be asleep on the basket in her lap and the two rough looking laborers stared with disinterest into space; but the young girl was alert and seemed anxious as she took in the sights and activities of the bustling area. We came to a multi-acre plot adjacent to the Navy base that had been cordoned off and construction was in progress.

The vehicles were pulled over to the side of the narrow street and we climbed out of the jeep to give instructions to the cement truck driver. The planned area had just enough room for a layout of four separate buildings of ten units each, in a square configuration along with a small community building and plaza in the center.

Of the four housing buildings planned for this site, one had been completed, and two more were under construction. I was amazed at how quickly the construction was completed once it was started. These forty dwellings were planned to be completed in two weeks.

As a side note, some of the occupants of houses like this were very creative, and would modify their structures for added comfort and amenities. Sometimes they would simply change the appearance to personalize the standard design. Returning later I would see, for example, intricately designed porches, beautiful paint jobs or even a second story added to individual units. I thought it was wonderful that they had the interest and desire to do this, and it certainly changed the appearance from the rather institutionalized look of the original houses. I was also impressed that most of the houses I visited later were well kept and neat as a pin. Whether it was the culture, the fact that the occupants were Navy men, or just the joy in having one's own abode in such a hostile war-torn environment, the units I witnessed were all well maintained.

Chu directed the driver to a location to unload. The ground was wet in spots after an earlier rain, but gradually the truck was backed in and placed for unloading in the least muddy area. Wooden planks were placed for transversing the mud when carrying the bags of cement to the mixing area.

As the crew prepared to unload the truck, Chu and I talked with Ming Cai. She stood eyeball-to-eyeball with me as she was a tall lady. She was all business in her heavy shoes, blue work shirt, and heavy blue serge trousers. She would have blended in easily on a construction site in the US. Ming was direct in her speech, "My crew will have the truck unloaded in about thirty minutes. Then, we'll return to bring the last load you ordered. I'll leave the old woman and the girl here; I'm told they have family in the area." Her English was excellent.

Chu and I walked over to inspect the recently completed building. A few head of livestock were roaming the streets, which was not unusual for this type of residential community near a military base. I noticed a couple of hogs and several geese as well as other fowl scratching around looking for food at the construction site. A large Tom Turkey in full fan regalia followed us for a while. We walked through the end unit on the finished building and noticed that the old woman from the truck was squatting in the kitchen with a small stove, burner lit, beginning to prepare a midday meal. Workers often used the new construction as a place to eat lunch and take breaks.

We made notes on progress of the construction and listed follow-up actions on the remaining cement work required for the two unfinished buildings.

After a while we returned to the truck to check progress with Ming Cai. We found her checking the count of cement bags as they were being unloaded by the two men from the shanty town. A work of drudgery, the men plodded along perfunctorily with bags they appeared to have been carrying for a hundred years. Then we watched as the young girl bounded onto the scene and began wrapping her head and gathering her hair tightly with the scarf she had brought. I was astonished to find her stride to the rear of the truck and grab a cement bag, hoist it to her shoulder and carry it balancing her way along the wet plank over the mud, barefoot of course, to the mixing area, some fifty feet away.

Now, I knew what those bags weighed. They were the same size and weight as an agricultural fertilizer bag that I used to carry from our truck to the corn planter on our farm when I was in high school. Those things were eighty pounds each. It didn't seem this girl weighed much more than that. And did she work fast! She was unloading nearly as much as the two men combined. It seemed so remarkable that I asked if I could get a picture as she did her job; she said she didn't mind. I took the picture and used it years later to impress upon my own kids the importance of hard work (and mightily bored they were!).

Since Ming Cai seemed to favor direct communication, I asked her a direct question, "How much do you pay that young woman?" She replied, "A thousand piasters a day." I thought a bit then said, "That's only fifty cents US!" She said, "That's right. She's the best worker I have, that's why she gets top wages. Ask me how much those two men get." So I did. "Seven-hundred piasters, but keep that under your hat please," she replied. She paused then said, "This isn't America here, Lieutenant, we are fighting a war and raising our families along the way. Good jobs, except for military, are hard to find, but I believe in being fair." I couldn't argue with her being fair, but I thought she was grandstanding a bit about raising families and all.

Just before Ming left for the last load of cement, she stopped to tell me that the old woman and the young girl had invited us to eat with them, in the housing unit where she was cooking on the floor. This surprised me since I didn't realize Ming Cai knew them that well. We thanked Ming and when she had gone, I talked to Chu, "What do you think, shall we take them up on lunch?" "It's a little unusual, but sure, why not; and besides I'm hungry," Chu replied.

Chu and I entered the house and found the old woman had placed clean decorated straw mats on the cement floor for us. She greeted us and invited us to sit. I thanked her in Vietnamese and made some small talk. We sat on the floor. Chu spoke to the woman and did a kind of introduction of all. He learned that the old woman was the mother of the girl. I was starting to feel guilty about taking from their meager food. I hoped that there was plenty and that the young girl would have plenty; the poor kid needed all the energy she could get with the yeoman-sized job she was doing.

Lunch turned out to be Vietnamese soup loaded with vegetables, most of which looked like kale of some type. Along with soup were rice cakes. The cakes were similar in taste to rice paper that I'd eaten before as part of a Vietnamese "sandwich." (Rice paper was very thin, and sometimes crispy. One would often see rice paper in the shape of a pizza dough spread over vegetation drying in the sun throughout Vietnam.) I had sneaked a look at the pot of soup boiling on the kitchen floor so I assumed the 212 degrees probably killed most of any living things.

About this time a young sailor from the Navy base appeared at the door carrying two bottles of 33 Export Vietnamese beer (in Vietnam they called it *Ba Mui Ba*, which means the number thirty-three in Vietnamese). He nodded at us and offered us the beer. The two women smiled at him and also nodded, and I got the impression that it was prearranged somehow. We accepted the beer and it was a welcome addition on this warm day.

As we ate lunch we talked and learned that the young sailor was the girl's husband, married only weeks, and we were eating in the unit that had been assigned to them from the Navy base. I thought the sailor seemed a bit light on seniority to receive the housing, but overall I thought it was good how things worked out sometimes. Lord knows they were dirt poor, she worked like a maniac and he was risking his life for his country. The lunch was good and I felt a little better about the whole project; maybe it would build some morale in the VNN after all … or maybe it was just the beer talking.

We spent the rest of the day taking care of business with vendors and materials requirements; and I began thinking about the young couple. I watched from time to time the young girl carry and mix cement for the slab they were pouring for the last building. I began to get the idea that maybe this was all just working out a little too well. In fact, I began to remember that I saw that young sailor talking with Ming Cai when we brought the truck in that morning. I also thought, "Chu is no slouch, he's a smart guy, I'm sure he's picked up on this, but sometimes my counterparts didn't tell me everything, like things I was better off not knowing."

I went over the sequence of events in my mind: Ming Cai won a contract from the US Navy to supply material to a VNN base; so she has a really good worker that she agrees to pay more than her other workers; so that worker is married to a sailor at the navy base where the project is located; so they need a place to live; so maybe she knows an officer at the base who works in the housing department; so she knows that Americans enjoy cold beer on a hot summer day. Yep, too much coincidence; must be a connection here. But then I thought, "What the hell, I'll just consider it as something that turned out well."

When Chu and I were finished that day, he had business in his office, so he headed to the shipyard in the jeep. Ming Cai offered me a ride back to my office on Hai Ba Trung in her old truck. Xao climbed in the old vehicle and Ming and I rode in the rear. We stood on the bed just behind the cab and held on to the railing above the cab's rear window. The breeze felt good on my face in the heat of the late afternoon.

We talked on the short trip up Hai Ba Trung Street. I asked her how she got started in this business. She explained, "My husband is deceased, I had three kids in school in Cholon, so I needed money. An associate and I bought the truck and began doing jobs hauling material, mostly government jobs. Then we found some workers who knew cement finishing, and the money was better. My associate found a job with the government, and I bought him out." I asked, "Do you have trouble getting jobs?" "No trouble," she answered, "but tax to the government is high, and I'm sure you know there is the corruption issue to siphon off a bit more."

We reached the OICC building, I said goodbye and jumped off the truck. The vehicle started up again slowly, and Ming yelled, "How was the *Ba Mui Ba*, Lieutenant?" I paused, then shouted back, "Cold!" "Good!" she said, "Next time we'll try Tsingtao, it's *Chinese* beer!" Yep, I knew she had made a deal.

We finished the Cuu Long project on time, and within a few days all units were filled. That was the last housing project I did with Chu.

A couple of weeks before I left Vietnam, I gave Chu a credenza that was behind my desk to use in his construction office to store records. This was part of the process of turning military property over to the Vietnamese. It was a nice piece of furniture that, in a time long passed, some exuberant government property clerk had ordered as a sort of symbol of the strength of America. Those aspirations were now long gone, of course, and we were trying to turn over all property post haste (with the façade of all legality intact, of course). Chu and Tran carted the credenza away one afternoon in Tran's Citroen 2CV.

I had promised Chu I'd stop by and say goodbye to his parents before I left Vietnam. When I went in to their tiny neat apartment one evening I was happy to see Chu there and we shared a drink. As I left, I couldn't help but notice the credenza sitting in the little entryway adorned with a beautiful bowl of orchids. Chu and I traded glances, and he looked down in embarrassment. It was an awkward moment for both of us.

He had given the credenza to his parents as furniture instead of using it in the shipyard as he'd promised. I shrugged it off and told him it looked great in the entryway. When I left Vietnam two weeks later he told me he was sorry about the credenza. I told him again to think nothing of it. Now when I think of the incident I'm amazed he even gave it a second thought at a time when his own future must have looked dark with the probability of Communist take-over once the Americans pulled out. In 1975 the new regime rounded up all junior officers such as Chu and imprisoned them in "re-education camps" for up to fifteen years. He may have gotten out of the country before then, but I doubt it.

Paul Yin

In January of 1973, hidden far back in the maze of streets that were Cholon, on the west side of the city of Saigon, there lived a Chinese man named Paul Yin. Paul Yin lived with his wife and three sons in a modest upstairs abode of four rooms on Hong Bang Street. There was a balcony at his front door facing the street that led to a stairway down to the entrance of his business on the street floor. There, Paul made the short journey each day for his work and when he pulled back the metal expander gate and unlocked the front door he stopped for a moment and admired the ornate sign that proclaimed his chosen profession.

Yin and Sons
Industrial Equipment Dealer, Ltd.

Paul Yin was the sole proprietor of a wholesale/retail equipment business that dealt exclusively in water pumps and plumbing supplies. The front of his store was quite small so that it belied the presence of a large warehouse nearly deep enough to reach the street behind his building. The warehouse consisted of four long rows of hardwood shelves, five tiers in height and spaced four feet apart, able to be accessed from either side. On these shelves, arranged in systematic fashion, were plumbing materials of all kinds, among which were copper piping, fittings, valves, faucets, unions, and water seals of all types and

We saw the small Datsun pick-up truck which bore the name "Yin and Sons" on the door.

sizes. Moreover, Yin's warehouse contained the most complete stock and selection of the most coveted and valuable plumbing item in all of Vietnam: the water pump. Each shelf space bore a prominent placard bearing the item part number, warehouse location and technical specification. I know all this because I walked along those long aisle-ways one day escorted by Mr. Yin searching for a water pump and studied those technical specs in detail. In the course of a couple of days I learned a little about water pumps and a lot about people from Paul Yin.

Cholon was literally a city within a city. It was Chinatown, Saigon. Historically, Cholon was the retrenched land and encampment to which the Chinese in Vietnam had retreated after the tables turned from dominating Vietnam over much of the first millennium. Now it was a bustling sub-city and a successful business and commerce center. There was no love lost between the Vietnamese and the Chinese, but they tolerated one another when mutual benefit was in play. Many of the older generation in Cholon never left the sub-city, and spoke the Vietnamese language only begrudgingly.

In January, '73 I was on a mission to find a water pump. I was working in those days with my Vietnamese counterparts at VN Navy bases along the rivers and the coastal areas. We were advisors and together with the Vietnamese sailors we put up buildings and piers for the boats, generator power stations, and water systems. We also constructed housing for the Vietnamese Navy men and their families, usually off-base, but close by.

In those modest little housing developments we usually built a small community center along with a fresh water system from the wells we drilled, and a sewage treatment facility. But because of the close proximity to the rivers or the sea, contaminated flowing ground water or salt water was a constant enemy in our search for fresh water for the houses.

I was not an expert in this area, but I learned while in Vietnam that the water beneath these bases was found in layered aquifers of ground water, salt water, and fresh water. The secret of obtaining good drinking water was in one's ability to accurately define these layers and then to design a good system to reach and pump the good water to the surface. The location of these layers and the distance required for pumping were important in defining the specific pump required.

It came about that we needed a very specific pump for houses we'd built for the families at a small support base named Ba Tri that was isolated along the coast. Bad water from the new wells was making the kids sick and livestock wouldn't drink.

Ba Tri was located in the Mekong River Delta at the point where the Han Giang tributary emptied into the South China Sea, a ripe location for a complex aquifer system. My Vietnamese counterpart for projects in the Delta was a civil engineer named Tran, and he knew something about aquifers. One day he and I and a French contractor named Antoine, whom I knew from the contract office in Saigon, met at the Ba Tri site. We took water samples, depth readings, and distance measurements in an effort to come up with a plan to fix the bad water problem. At the end of the day we concluded we could make this system work with a new well and a special large pump. Antoine told us from his experience, that if one needed a special pump in Vietnam, the man to see was Paul Yin in Cholon.

Antoine also said to expect to pay a premium for the pump, for Yin was a hard-nosed businessman and he seldom yielded much on price. He usually held most of the cards in a pump crisis, and he knew it. We didn't have much choice, Yin was about the only place to get the right pump in any reasonable amount of time, including the US Navy Supply System. At this point in Vietnam, few in the US Government were willing to send much material to Vietnam after the billions poured in there over the previous decade. They were interested in turning the equipment that was in Vietnam over to the Vietnamese, and getting out; certainly not sending more.

So we headed to Yin and Sons to pick out a pump. Antoine drove us in his blue Land Rover bearing his company's Mecanique logo on the doors. We pulled up on Hong Bong Street and parked the Land Rover under the Yin and Sons sign. After some discussion in the small front room with a lady whom we assumed was Mrs. Yin, we learned that Yin preferred to speak only in his Chinese dialect ... no English, no Vietnamese. But she told us in Vietnamese that her eldest son often translated Vietnamese to Chinese and a little English to Chinese for his father.

Yin entered the room with his hand on the shoulder of his eldest son, pushing him just slightly forward; Yin was wearing a broad smile and nodding vigorously, and repeating something in broken English. He was short in stature and a rather rotund man with a bristled black flat-top and bright eyes behind black-rimmed glasses. His movements were quick and he seemed energized from the moment he entered the room. Quite contrary to the stern businessman I'd expected, he appeared friendly and forthcoming, and the kind of person who was always on a mission. Yin was definitely a man of action.

I nodded to him and tried to introduce our party with some Chinese phrases I had practiced the night before; phrases like, "Good morning, Mr. Yin, my name is so-and-so, this is Mr. so-and-so," but it didn't work. He kept smiling at me and was focused on my uniform, and then I realized he was chuckling and saying in broken English in a rather exuberant manner, "American, American Navy!" Yin seemed impressed that the American Navy had come to do business with him in his shop. It seemed to bode well for us at that point.

"Yes! Yes, of course; American Navy!" I responded and tried to engage him, but quickly found he really knew almost no English. But we began to set up what became an effective translation line. On the one end I addressed Tran in English and broken Vietnamese, who in turn translated in Vietnamese to Yin's number one son, who passed on the message to Yin in the local Chinese dialect. It began cumbersome but was soon humming back and forth. I must say Yin really got into it and was shouting answers and return questions to his son, while occasionally chuckling and being quite animated.

As we went back and forth through the line of communication, Yin would shake or nod his head and speak staccato Chinese, but I noticed he was also prone to throw in among the Chinese some American marketing slang phrases he must have picked up in newspapers and television ads. I wasn't too sure he knew what they meant. There were words like, "Good deal! Two-for-one! No charge!" And, "Special Sale!" I caught the eye of the Frenchman who had also picked up on the slang and we shared an amused smile. The negotiation continued with constant jabbering up and down the line in all three languages simultaneously. Antoine and Yin's wife stood back and quite enjoyed the show.

Technical talk was the easiest, a common language among all; terms such as total dynamic head, horsepower, and gallons-per-minute were understood and agreed upon. It wasn't long before we learned Yin had the very model pump we needed somewhere in that warehouse. It was an industrial Mitsubishi unit and we proceeded to verify the exact part number. A shot of adrenaline was then injected into the fray, and both sides became motivated to settle on price. The buzz in the communication line grew to an elevated level, faster came the comments, and more exaggerated the animations. But if the buzz was high, so was the price, hardly moving from the original inflated one suggested by Yin.

We were at a stalemate after a few minutes, and so to mix things up I harkened back to one of Yin's slangs and shouted, "No good deal!" This needed no translation and everything suddenly went silent. The smile left Yin's face as he ushered his son off to the side and asked for a small spiral-ring note booklet which his son kept in his shirt pocket. Yin turned and looked at me sternly as though to admonish me for an arrogant outburst. At that moment the other two Yin sons entered the room with rather foreboding expressions. For a moment, our party traded glances half wondering if we were going to have to fight our way out of there. In silence, Yin studied the numbers in the booklet for a few seconds and then slowly a smile began to creep across his face. He spoke some orders earnestly to his son who began translating immediately. But even as his son continued, Yin could not contain himself and broke into a broad smile and cried, "American Navy! Good deal! Good deal! American Navy!"

When the new numbers were finally translated, we realized he had reduced his price only incrementally, but we decided this was probably the best we were going to do with him. I must say that at this point we were still a little leery of exiting the room intact, so any reduction in price was just gravy. Maybe it was also that his laughter and enthusiasm were contagious and soon all in the room were shouting and laughing "Good Deal!" But what was really absurd was when, after the noise died down, Yin cried out loudly, "Pocket the savings! Yes! Pocket the Savings! American Navy! Ahh! Pocket the savings! This was really too much, and was such an unexpected and ridiculous comment that the whole group lost it and broke into out-of-control laughter.

Yin then hurried to a cabinet at the side of the room, and produced some small green heavy drink glasses as his wife poured rice wine and began distributing the drinks around the room, as if to seal the deal. Yin shouted the toast: "American Navy! American Navy! Pocket the savings!" And then downed the wine in a flash. It was such an absurd scene that one could not help but laugh out loud and follow his lead. We knew we paid too much for the pump, but at least it was entertaining.

After all the hoopla, Yin barked some orders to his son and we followed him into the warehouse in the room behind. Here, Yin showed off his impressively stocked shelves as his son retrieved the pump for our inspection. I remember the pump was a magnificent piece of equipment, dark blue in color and heavy enough to require two men for transport. The silver name plate was emblazoned with the name Mitsubishi and the technical specifications. Then he retrieved manuals and had the pump packaged severely so as to protect it for the trip to Ba Tri. While the pump was being packaged, Yin quizzed us regarding details about the use of the pump and location of the installation and about the system itself, especially the piping size and type, the type of filters we were using and the surrounding equipment. As he digested the information about the installation he grew rather thoughtful but said nothing.

As we walked out to the Land Rover, Antoine said, "Man, did he have that staged; he played us like a violin." I agreed, "He was good. He hit all the basic tools … enticement, flattery, unspoken physical threats, and even alcoholic reward! He fleeced us and left us happy. Now that's talent!" The packaged pump was loaded into the Land Rover and we bid the Yin family good-day.

* * * * *

Next morning Tran and I loaded the pump into the bed of his tiny Citroen 2CV truck and drove south out of Saigon, crossed some tributaries of the Mekong and on to Ba Tri. There we met the Vietnamese contractor who we'd worked with building the row houses and the water treatment system. He had finished drilling and laying the pipe to the part of the aquifer we had specified and was ready for the pump.

We went to work on the installation of the pump immediately. Previously the men at the base had poured a cement slab to mount the pumping equipment and by noon the pump and piping had been installed and the electrical hook-ups completed. It was time for lunch break but everyone was so anxious to see if the system worked that we decided to fire it up before eating. Several residents of the new houses and many kids were standing around in great anticipation as we lit it off. We hit the switch and wham! The equipment vibrated a bit at first and then smoothed out into a moderate hum. Water began spitting and surging into the cistern. The clear cold water looked and smelled good and all were elated, but the spitting didn't stop and elation turned to disappointment when we realized that only a small amount of water was sputtering out while the pump whirled at full speed. We left somewhat dejected to go eat lunch, and think it over.

When we returned from lunch a new set of tire tracks told us a vehicle had driven down the muddy grass path that connected the housing area to the base. As we continued we saw the tailgate of a small white Datsun pick-up truck that had parked there to avoid the muddy area around the new construction. We approached the truck and saw that the door of the truck bore the name of Yin and Sons with the Cholon-Saigon phone exchange below.

"What the heck is he doing here?" I asked Tran. Tran shrugged his shoulders, "Nothing would surprise me about Yin after yesterday." We walked to the pump site where Yin was barking orders to his two sons that were with him and to a worker from the contractor who had helped us install the pump before lunch. There were plumbing parts and short pieces of pipe of different diameters laying about, along with a number of tools. I looked at the set-up and realized that it was changed from the way we'd left it before lunch. Yin came hurrying up to me when he saw us and grinning all the while he launched into an enthusiastic couple of minutes of pure Chinese with animated hand gestures pointing out what appeared to be new parts of the system.

I asked Tran if he could find out from Yin's eldest son what was going on. In Vietnamese and broken English his son explained, "My father said this pump will not work properly unless the system is balanced. It will cavitate and ruin the pump. My father said it requires pressure reducers in the pipes and a pressure accumulator." Yin was vigorously nodding his head with a broad smile all the while his son explained what was happening, and his other son finished the modifications. Then Yin shouted more commands to his sons and the pump jumped to life spewing a heavy stream of clear water into the cistern. "Ahh!" Yin laughed heartily and led the crowd in clapping and cheering. He shouted a stream of Chinese to his sons, us, and the crowd in his enthusiastic manner while he pointed at the system and the water flowing. I told Tran that you have to admit this is pretty impressive, but I also wondered how much this was going to set us back. We were already way over budget.

Yin's sons put the tools back in their pick-up truck and the three of them prepared to leave. We walked over to the truck and thanked Yin for coming down and fixing the installation with the new parts. Then I asked what the charge would be. Yin of course did not understand the question so he turned to his son who began to explain. Yin cut him off almost immediately, and shot out some words in Chinese toward us with a big smile. Then as he reversed the truck to turn around, he leaned out of the truck window laughing heartily and shouted, "No charge! Pocket the savings!"

Images

It was a beat-up black '66 Chevy four-door Navy sedan, three-speed stick on the column and black walls. We trekked out of the Duc Sieu Hotel every morning that we were in town, always in the dark, usually about three or four of us. We nodded to the Chinese rent-a-guards as we passed the guard shack, jumped into the Chevy and began the twenty minute ride in the dark. Never did our small elderly Vietnamese driver in dark trousers and white short-sleeve shirt ever turn on the vehicle headlights. He always felt his way using the sparse street lighting and residual light from the few taxis which operated that early in the morning. I felt he drove too fast for the lighting conditions and from his viewing position which was just below the steering wheel arc, but he never seemed to have a problem.

Not much traffic on the streets at that time of morning as we careened down Vo Tanh Street, then turned north for some blocks past the Splendid Hotel, then left by the Cathedral and the John F. Kennedy Memorial Park on over to Trung Sisters Street and whipped into the OICC garage. We crawled out of the car and headed up the concrete stairwell to our OICC office in Saigon.

I always liked this time of day, and the ride over to Trung Sisters Street. I think it was the coolness of the air and the feeling of flying through the dark on nearly empty streets as night still lingered. Un-

The kids began to gather around her and chant, *"Em Mi, em Mi!"* ("American kid, American kid!")

usual scenes, certainly foreign to most Americans, would come in and out of focus as we sped along. Forms like that of a lone pedi-cab driver cycling along the edge of the street with a flashlight lantern and a polio-withered leg, or a young secretary off to work early in a pristine white flowing *ao dai*, motoring along on her Lambretta. We kept the car windows open as we raced through the dark city. The damp air brought the pungent smells of Asia to each as we sat in silence and observed, and thought about the day ahead. I was thinking about my business that day going down to the huge Saigon seaport of Newport and later to a town called Phu Cuong.

Jim Landers beat me to the office that morning. He threw a stack of NIS sheets on my desk from the overnight bag. Jim was a Navy Warrant Officer about thirty years my senior, who had come up through the enlisted ranks. I developed a great friendship with Jim during the time I was in Saigon. He was a gentlemanly father figure to most of us twenty-somethings. Jim had a quiet manner and a big smile, and his savvy advice helped a lot of us through the year over there. He had a son in his twenties who was in the Navy stationed in Japan. Jim was the Division Officer of the dozen or so enlisted Seabees attached to our outfit. Jim was a model of personal behavior: brave, friendly and helpful.

"Some activity around An Loc last night; where are you jokers heading today?" Jim asked as he pointed to the sheets on my desk. We received intelligence reports each day from Navy Intelligence Service, but usually they were not much help as we traveled around South Vietnam, mainly because they were always after the fact. They contained raw data that read something like: Four VC spotted along trail two clicks southeast of Ben Luc carrying mortar tubes, or two M37 weapon carriers blown up by boobie traps one click from Navy Base at Nam Can.

"Tran and I are heading north, up to Phu Cuong with a bunch of trucks with building materials that are supposed to be loaded at Newport. Want to join us?" I said. "No way, José; got a date." "You dog," I returned, "What kind of date?" "The Air Force Major kind, out at Tan Son Nhut. Remember I'm heading home in two weeks, I gotta get my flight and orders verified by the Major. I'm getting so short I could sit on a razor blade and dangle my legs." "You dog," I said again.

At 10:00 A.M. I threw my bag into my counterpart Tran's tiny Citroen 2CV truck and we headed through heavy diesel-filled traffic up the road to the Port of Newport. It was the new facility the US had built in 1967 just north of the old Port of Saigon which wasn't large enough to handle all the cargo for the US war machine. At 10:45 we pulled into the Port and started looking for a freighter hopefully delivering the building materials we'd ordered weeks ago. Tran and I needed the material for the construction we were organizing at a Vietnamese warehouse facility at Phu Cuong. The port was jammed with trucks, people, noise, and activity. Temperatures were beginning to rise as people scurried about moving material in and out of the port.

We reached the piers and found that the large freighter had docked late the previous day. I recognized the vessel from the tracking reports we'd been monitoring. It was a big relief to see her there since I had hired eleven semi-trucks from a Vietnamese contractor to haul the building materials up to Phu Cuong. The semis were due at noon, not much time to get the material inventoried and received quickly to be ready when the trucks arrived.

Sprawling ports of entry are confusing places to figure out who's in charge and what it takes to receive material. With the help of Tran and his Vietnamese, we finally found the Port Master and learned the process, part of which was me climbing down into the main hold of the ship. That's when I remembered I had a touch of vertigo and wasn't at all sure I could inch over the side of the forty-foot steel cliff into the huge cavern, and climb down to the bottom on the widely-spaced rusty steel rungs that had been welded to the side of the hold ages ago. Pure peer pressure motivated me into action when the Port Master's assistant said "You need to identify and inventory that material in the ship, follow me to the bottom!"

While down there I spent more time worrying about going back up the steel ladder than taking inventory, but at last completed the task. Peer pressure struck again to crawl skyward when the swarthy shoremen shouted, "Go on up, we'll follow after you!" The trucks began showing up as I inched my way to the top and pulled myself over onto the flat steel decking. Tran was already directing the trucks into position and soon the loading began.

After all the material was accounted for on the vehicles and all documents signed, the caravan of eleven semis with our little 2CV leading the way headed north for the short trip to Phu Cuong. We made several stops at check points and once at a bridge on the way. Tran made certain the caravan stopped at the bridge. "What's this all about?" I asked Tran. "Just a little problem like that sign there that says: Stop or Get Shot!" he replied. "Good call," I said.

As we approached Phu Cuong, the scenery became magnificent. There were beautiful palm groves along a boulevard south of the small city, the air was clear and the breeze carried a hint of the sea. In the distance a range of low-level mountains rose to accept a cloudless blue sky. We were encompassed by lush vegetation an all sides. I mentioned to Tran what a fantastic place to have a home. Tran replied, "It's beautiful, but then there are the tunnels somewhere below." The tunnels I learned later were the below-ground infiltration routes the VC and NVA used to move war materials south. And they were all around the area of Cu Chi close by.

Arriving at the construction site, the drivers pulled their trucks among tall palm trees and the abundant lush vegetation on the outskirts of the city. They interwove the long trailers until all found a parking place from which their load could be accessed while still capturing shade. The trucks were shut down and the drivers gathered and chatted in a good natured manner as they found an open area to eat their midday "brown bag" meal. After lunch they tied hammocks under their trucks and took afternoon siestas.

Tran and I found sustenance in a little stand where some Vietnamese women were selling what they called spring wraps on top of a plate of rice. I bought two plates and two iced teas for Tran and I, and we sat down in the shade with the Vietnamese construction manager to go over the plans and the site layout.

Some kids, perhaps a dozen or so of kindergarten and elementary school age, were playing nearby in an area adjacent to a small school, laughing, and making lots of noise as kids do. They recognized that I was an American, and sometimes they would go running by as a crowd and shout to me and laugh, *"Ong Mi! Ong Mi!"* (American! American!) to see if they could get a reaction. To tease them, I pretended to ignore them for a while, then later shouted back to them and returned their laughter. This delighted them, as they screamed and laughed the louder. At length, they tired of the game and began to play quietly in a small grassy yard nearby.

Tran and I continued to review building plans with the site manager when I suddenly became aware of one of the kids standing behind me, perhaps fascinated with the engineering drawings. But she surprised me when she said in a tiny little voice, *"Toi la em Mi"* (I'm an American). I turned and was astonished to find the Vietnamese voice came from a small girl of perhaps four or five years old, with nut brown hair and blue eyes. Now that was something you didn't see everyday. She could have been a kid in any American playground ... until she opened her mouth.

Some of the other kids came and began to chant around her, *"Em Mi! Em Mi!"* (American kid! American kid!). The child of an American GI and a Vietnamese mother. She was smiling, and the kids, at this age, were laughing mostly with her, but it was obvious some of the older ones were making fun of her. She had not yet grasped her predicament in life, but would soon understand, in years that would come too quickly. Even at that young age, the little girl was beginning to be singled out from her school mates. I asked her in Vietnamese if she liked the drawings. She just stared. Then her young mother appeared, took her by the hand, and led her quickly away, glancing back toward me and looking down with an embarrassed smile.

Much has been written about the children of GIs and Vietnamese women. Most were left with no fathers, and in some cases, no mothers. The children were relegated to a class such as that reserved for lepers and called *"Bui Doi"* (children of the dust). A few were reunited with their fathers in the US or were adopted by American families. Others were taken in by orphanages and charitable organizations within Vietnam. But most lived lives ostracized by their fellow countrymen for sins they didn't commit. With no tools to resist, most subsisted among the lowest of the low as beggars, prostitutes, or street people selling cigarettes and toiletries in the big cities.

The truckers finished unloading the material by 4:00 P.M., and then Tran and I headed back to Saigon. Tran dropped me off at the OICC building and I met Jim Landers for dinner at our mess hall.

It had been a long hot day, and as it finally wore into dusk, Jim Landers and I took a colleague he'd been meeting with back to the Meyerkord Hotel, a multi-story downtown hotel converted into quarters for US military personnel. We pulled up to let Jim's colleague out, and Jim went in to get us a couple of cups of ice cream which the USO sold in the lobby sometimes. I kept the jeep running. A man in his twenties came up to my window holding a note pinched between his two wrists. I say wrists, because he didn't have hands. He kept repeating the words, "Give me money, *Dai Uy* (Lieutenant)." The note was printed in English and announced he was a veteran and legitimate war beggar; it was laminated with well worn rolled-up corners. He had a small plastic bag pinned to his shirt; I stuffed some piaster notes into it trying to be alert to devious intentions. There were none.

In the meantime, a slight girl of perhaps sixteen or seventeen crawled into the jeep from the passenger door Jim had left ajar, and sat next to me. She wore a filthy pink tee-shirt that bore the faded image of Chubby Checker surrounded by the words, DO THE TWIST! The longish tee-shirt was the extent of her apparel. She was shoeless and listless and kept saying, "Take me home tonight, *Dai Uy.*" Jim returned with two cups of ice cream, and, assessing the situation, handed them to me. He carefully picked up the little girl and carried her over to the sidewalk. He put her on a rusted wrought iron bench, then came back for his ice cream. He handed it to the girl which she devoured. He later said she was light as a feather.

On the way back through the city that night we passed the Chinese Embassy, and Jim said, "Hey, I forgot to tell you, I saw Kissinger today. He was going into the Embassy there in one of those black State Department cars." "What was he doing there?" I asked. "It's the China thing, China is the new thing, man. Taiwan is out, Mainland is in. Nixon wants to move on. He wants out of Vietnam. He wants to open China!"

Jim dropped me off back at the Duc Sieu Hotel. The old converted hotel once housed dozens of Navy engineering officers, but was now soon to be closed. In the summer of 1972, with the Americans now leaving Vietnam, I was one of only a few left in this large old structure. My roommate, Jack Barrett, had recently moved out and I was now alone in the room. Jack had his own problems of trying to get his Vietnamese girlfriend out of the country and was staying with a friend near the American Embassy as he processed paperwork the last week he was in country.

I turned in and fell asleep immediately, but was awakened by something around 2:00 A.M. I felt my way through the darkness out into the main room. I was told the Duc Sieu was formerly a French Hotel, perhaps built in the early 1900s, and leased by the US Navy. As a French colony, the style back then was that of full service designed for comfort to which the French were accustomed. I recall the rooms were large and the hallways and passageways wide. Ceilings were high and tile was employed liberally. I was on the third floor and the exterior wall of our main room, which looked out the front of the hotel, was floor-to-ceiling windows with a French door that opened out onto a balcony the length of the room. At two in the morning, the balcony itself was dark but the lights of Saigon illuminated the cityscape beyond. I walked out onto the balcony.

The shadows on the balcony gave me cover and hence the freedom to gaze out over and across the city at night, unseen and yet seeing. Some nights, if I couldn't sleep, I would sit on the balcony in the dark with a beer and just listen and watch the city.

This was one of those nights. I could make out the lights of some of the downtown bars with their neon signs and faint laughter. A few taxis were still circling about. The city on the whole was closing down for the night and the sounds coming from across the spectrum below were not the din of the pulsating city during the day, but individual sounds that would echo between the dark buildings. The scratchy exhaust note of a Blue and Yellow Co. taxi, long overdue for a tune-up, stood out clearly. A musical note struck by a wrench dropped on the street, a person coughed on a verandah somewhere within the block. I could hear undertones of the night shift conversations of our Chinese guards below, not ten meters away. A dim light was on in the guard shack where they were playing their games of chance. Thunder or artillery in the northern outskirts of the city. Some dogs barked on Vo Thanh Street two blocks away.

I looked out over the city and could feel a coolness in the air beginning to creep in. The warm, thick air that required effort to draw in and then push out in long deep breaths was being replaced at last. As the city cooled, it would gradually transition into a light slumber, but never a deep sleep. Saigon would take its rest with one eye open. There would be perhaps two or three hours of such a state before the early risers would begin to stir again on the streets. Then the whole scene would repeat and another day would be added to the more than decade of a sort of artificial existence and balance that had become routine since we Americans had been here. Much of the population seemed to have gotten used to the routine and hoped it would sustain. But the reality was it was a stalemate supported by the mighty American military on the one hand and the Communist backers of Ho Chi Minh on the other.

With American forces drawing down rapidly, I wondered how many more times these artificial days would repeat. It was a question beginning to be debated privately at all levels as the country watched for signs from Henry Kissinger and Le Duc Tho, the power brokers representing America and North Vietnam respectively, as they negotiated and bargained the fate of The Republic of Vietnam. A nervous undertone had formed beneath the daily routine across the cities and countryside, while people tried to foresee how much longer the status quo would continue and where the dice would fall when it was over. Contingency plans were being made, often in secret, as the wise or those with wealth maneuvered for positions or potential escape routes. Even my own counterpart Tran had asked me for help in mailing a packet to a Swiss bank address. He didn't disclose the contents, but I'm guessing it was financial instruments in connection with a potential life after Vietnam if everything went sideways.

A cool breeze swept up to the balcony and it felt good as it helped to push out the last of the oppressive heat. Images of the day bounced around in my head. Visual sound bites as it were, reappeared as I recalled events that would stay with me for years to come: a man with no hands, or the hungry girl in the jeep, or the Amerasian kid at Phu Cuong.

Tomorrow there would be more sound bites for sure as I traveled to the Vietnamese Navy Base at Nam Can, in a clearing along a river in the thick U Minh forest in the southernmost part of South Vietnam, on the Ca Mau Peninsula. There would be many more days of images that would stay with me but the question this day was how many more. I turned in and set the alarm in time to catch the 1966 black Chevy to Trung Sisters Street.

Crippled Caribou

The plane was called a C7 Caribou. Named after an animal. It was a smallish cargo plane that was used in Vietnam by the US Army and Air Force. There were two props driven by Pratt Whitney Twin Wasp engines. The plane was a DeHaviland, and I'm guessing the Caribou moniker was the result of it being built in Canada. DeHaviland also built a plane called a Buffalo, and one called an Otter.

Being a cargo plane, ferrying people around was a collateral duty. The seats in these things were in the cargo bay, of course, and were the definition of utilitarian, "beaucoup" uncomfortable. Very hard with no give. They were long fold-up bench seats which lined both sides of the fuselage, made of heavy webbing suspended on some aluminum tubular structure, and reminded me of those old Army cots that they sold tons of in the '50s and '60s. The floor of this plane was all metal with lots of tie-downs for cargo. I noticed there were a few pallets of material secured to the floor toward the rear of the cargo space. They looked to be some type of vehicle parts like axles or such. There was also a pallet of lighter equipment that included two Samsonite suitcases and speakers or sound equipment of some type that were covered with some plastic wrap to keep moisture off; presumably.

The airplane had two spindly-looking long landing gear struts that were anchored to the rather spindly-looking high wings. The

This time we knew we were landing, one way or the other.

fuselage hung below the wings so, when landed, it was low to the ground and easy to load.

Everyone on the aircraft but the crew was hitchhiking. I sat with my back against the fuselage, shoulder-to-shoulder with a bunch of people in uniform, none of whom I knew. We were looking across the floor of the airplane into the eyes of a bunch of people looking back at us. The one exception to the shoulder-to-shoulder arrangement was a slight gap in the area where a six-inch red band had been painted around the interior of the fuselage. The red band denoted the area of the rotating propellers. If a propeller should come apart for any reason, parts could be flung through the fuselage and anyone in that area would become instant dog meat. Most people, no matter how macho, shied away from that zone.

There was no verbal communication because of the high noise level. The Caribou was no speedster, and we bounced along like an old wagon crossing the plains in this very hot and very noisy airplane. One thing I observed about being in an environment of high decibels, vibration, and heat is that one loses his normal social inhibitions. With verbal communication impossible one begins to get drowsy and personal space often becomes invaded with no consequence. That is to say it would not be uncommon to observe one riding in a semi-conscious state staring directly at another across the airplane whilst they stared in return at you slouched against a neighbor who may actually be asleep on one's shoulder.

We had been flying up there in the clouds and at a high enough elevation where it was more difficult for unfriendly people who wanted to do bad things to us, like shoot us down, to do that. They would have to really plan and work at it, whereas at the lower altitudes we became visible and more available to diabolical individuals. The lower altitudes usually increased the tension in the aircraft.

Flying and jostling along I began to anticipate that tense period as we approached Tan Son Nhut Airbase in Saigon. The airplane swung out over the Pacific and banked, engine speed changed, and new noises were introduced. The landing gear came down with an abrupt bang accompanied by abrupt deceleration of the airplane with those long landing gears now hanging in the wind. Wind, mechanical, and hydraulic noises were all normal; passengers woke up and were looking around as we entered into the tense period where each is privately thinking, "Good, we've got about twenty seconds to get through and we'll be home free. Once again engineering and technology and good old 'Yankee know-how' (or should I say 'Canadian know-how' in this case) will have triumphed over the forces of evil!" I tried to turn around and look out one of the few windows behind me to see where we were, but none was in my vicinity. But I could see a portion of the right landing gear through the window across from me just above the heads of the passengers. It didn't look right.

As we braced for a shift forward anticipating the brakes when we hit ground, I began to notice that three young individuals sitting across from me were definitely not in uniform. They seemed to be native Asians and were wearing '70s style "Mod" clothes. There was a young lady flanked by two young men.

Just as we expected the wheels to touch down, the pilot gunned the engines to full throttle and pointed the nose skyward. Bracing for the opposite effect, we all fell into our neighbor to the rear of us. Not good. Not normal. A minute later, the young co-pilot came to the cabin door and stood hanging on during the steep climb and shouted as loud as he could to us, "Right landing gear won't lock down ... it's stuck, we're going up for a while and try to shake it loose!"

Several bad things came to mind, "It's that danged altitude thing again! OK, we're going to fly around in the sky at fairly low altitude and do our best to attract attention and act as bait for people who are rewarded by their comrades for shooting down an American plane; that's just great! Second, this airplane has been flying for awhile, how much fuel do we have to screw around with up there, and what if they can't knock it free, you can't exactly put a wrench on the errant part; it's out there, and I mean way out there, on the wing. And most importantly, it's becoming really hot and stuffy in here and I'm not feeling too good all of a sudden." I don't get along well with roller coasters. I can barely handle the scrambler. I tried to remember what I had for lunch seven hours ago.

We flew in a circle just below the clouds, first in a steep banking maneuver then in a series of rollercoaster ups and downs that were free-fall then buoyant acceleration, free-fall, back up, and so on.

I looked across the plane and it looked like the landing gear may have budged a little. Then I looked below the window and didn't like what I saw. The three Asians were in bad shape. One had his hand over his mouth, the other young man was trying to console the girl who was turning green and had pulled out a silk handkerchief and had it over her mouth; her eyes were closed.

Please do not barf ... please! Whoops, too late, damn! Next would come a real life example of scientific gas diffusion we learned in chemistry class, which would spread the good news throughout the already stuffy cabin. Desperately I tried to focus on icebergs, waterfalls, anything.

I started feeling the old familiar wave of something rising but managed to hold it in check. The co-pilot came back and shouted, "We'll try again, we'll fly low over the tower first to see what they think." Down we went, then banked over the ocean and came in low at speed over the tower. A few seconds, then full power to the below cloud altitude again. All three of the Asians hurled during that process. Crap! I was just thinking that I could handle fear, but fear and that unwelcome wafting together, now that was a lot to ask. I was beginning to feel really bad. I breathed through my shirtsleeve to help keep the demons at bay. It helped a little. I looked down our bench, two of the soldiers had now tossed cookies, I thought, "Damn, this scene is getting worse fast!"

We tossed and banked and dived another five minutes up there to try to break that danged landing gear loose, then the plane leveled out for a few minutes as it headed out over the ocean again. The co-pilot appeared again in the cockpit door this time eating a Clark bar, with the orange wrapper hanging. Three Army guys puked immediately. "We'll try again, third time's a charm!" he yelled, smiling. Here we go, we banked again, then came in fast and low and hung forever at about fifty feet off the runway. Then max power and up we go again, skyward. Third time was no charm after all! Who in the hell was the moron that made up that stupid saying anyway! Third time was worse than the first and second combined. Passengers were now moaning and in actual pain and holding their stomachs as most yielded to instincts and no longer fought the rising tide.

The airplane was a mess, and for the first time I was beginning to think we were going to have to land on one landing gear; I was trying to think what that would be like. I thought I remembered they would dump most of the fuel first before they made the attempt. I was thinking this over, breathing through my shirtsleeve, when the co-pilot entered the cabin with a heavy crank-shaped tool. He worked his way to a point adjacent the right landing gear and began to open an access door which was right between the young lady and one of the men. "Not feelin' too good, huh?" he shouted to them.

They didn't answer. Having opened the access door, he fitted the tool into a socket and began applying cranking force left and right to no avail. Two beefy Army guys who had maintained their composure came to his aid, and following the crewman's instruction, the three of them broke the gear free.

They cranked the gear partway up, then all three together cranked the gear down as far as it would go. The gear appeared to be all the way down, but it was not locked. "There's a back-up plan for a busted lock!" he shouted to his two helpers and smiled as he pulled a steel pin from his pocket and pinned the crank to the socket. Then he slipped a wire rope cable-end over the crank and winched it tight to a spar at the edge of the cargo bay. He turned and shouted to those watching, "That oughta do it!" He helped one of the civilians lift the young lady over to another seat and crammed her in between two others. She was in total disarray, her head hung down and her green face was partially obscured by her long black matted hair partially stuck to the side of her head. Then he ordered the area around the band-aided gear cleared, and room was made for those passengers elsewhere along both sides of the fuselage. He took a seat near the gear, picked up his head-set, and gave the signal to the pilot.

At this point the cabin was a disaster; it was boiling in there and the air nearly unbreathable. I was beginning to wonder if passengers would start passing out due to lack of oxygen. No one was complaining, though, as we entered the landing pattern out over the Pacific, banking once more, leveling out and dropping in altitude. This time we knew we were landing, one way or the other; no one breathed. We hung in the air a few feet off the runway forever, then a touch on the asphalt, a slight bounce, and then down and the wheels stuck, the gear held and we coasted to a smooth stop. Suddenly smiles all around and scrambling to get out of that cabin and off that plane.

Jim Landers met me in the gray navy jeep. I climbed in and he said, "Stay on your side, man, you stink!" We headed back to the office on Hai Ba Trung. I got cleaned up and met Jim at the mess hall later. Jim said he got bored watching us go up and down in the Caribou. I told him I didn't want to discuss it before eating. After dinner Jim said there was a band at the USO downtown on Nguyen Hue Boulevard later that evening. We agreed to walk down in a couple of hours and see if they were any good. At around 8:00 P.M. we sauntered down to the USO and sat down with a couple of 33 Export beers. The band was better than I expected; American rock and roll, and country music as well. They turned out to be a Filipino band hired to perform shows for G.I.s throughout South Vietnam.

They sounded good in the old French building downtown that echoed with their drums and guitars. The music sounded so good that we overlooked the rather pronounced Filipino accent. The lead singer was a young lady with a wonderful strong, voice.

But there was something familiar about her as I listened to her sing and watched the show and I thought maybe she had been on an album cover or such. She was wearing a gold sequined dress and heels and had long beautiful black hair. At the break, the band came walking toward us as they left the hall for a drink. The lead singer had her eyes fixed on Jim and I. Following behind her were two young men who had been playing guitar and drums. As I watched her approaching us I thought to myself, "No way, can't be."

When she passed us she smiled and said, "Did you like the plane ride, Lieutenant?" Jim looked over at me and said, "What the hell went on in that plane, anyway?!"

Saigon Christmas

It was Christmas Eve, 1972. I stood in a shadow back from the street, and watched the pedestrian traffic pass below the gaudy Christmas lights while I waited for the young boy to show. I was in Saigon and finishing up a year in Vietnam as a Navy advisor. The war was winding down and Americans were leaving; I had expected to go home early; certainly way before Christmas, but it wasn't to happen. I eventually finished my tour as planned and left in January '73. I'm glad I didn't leave early, though, because it gave me a Christmas Eve I'll never forget.

One of my jobs over there was advisor for the Vietnamese Construction Bureau. We were building houses for Vietnamese Navy families along the rivers mostly in the Mekong Delta. I traveled with my Vietnamese counterparts, but was now living in Navy-rented quarters on Hai Ba Trung (Trung Sisters Street) in Saigon. There were a few other Navy officers living there and we had a Navy mess where we ate most of the time when we were in town. The Navy hired local Vietnamese to prepare and serve our meals in the mess. Since I took most of my meals there, I got to know most of the staff on a friendly basis.

In Vietnam I knew a handful of Americans who married Vietnamese girls and took them back to the States. One day, I received a letter from a Navy Officer in Gulfport, Mississippi named Tim

The old woman sat side-saddle on the rear of the ancient motorcycle.

Brockman who was trying to get his girlfriend back to the States, and asked me to help with the paperwork. It turned out she worked in our mess hall … a server named Lana. She was very young, perhaps nineteen; a friendly girl but very quiet. She spoke no English, but I could converse with her a little in the limited Vietnamese I had picked up in the required language school for advisors.

I later learned the Vietnamese servers had recommended me to Brockman as an American likely to help Lana, after they learned I was going on R&R to meet my wife in Hawaii. I grabbed a Navy legal pad and scratched out a letter back to Brockman saying I'd try to help them. I synchronized Lana's flights with mine to Hawaii and helped her prepare documents. I sent the itinerary information to Brockman so he could meet her at the airport in Gulfport.

The trip went off without a hitch except for a tearful farewell with Lana's mother, her younger sisters, and a little brother at Tan Son Nhut Airbase. It was obvious that the family was living on the edge. Their clothes showed their hard-scrabble life, and the sadness of Lana's leaving seemed to hang over the little family and made the scene all the more pitiful. They viewed this parting as the last time they would ever see Lana.

In Hawaii, I helped her through passport control and customs, showed her the way to her connecting Atlanta flight and wished her good luck. She thanked me and walked away down the concourse at Honolulu, scared but brave. That's the last I saw her, but I know she made it safely because when I returned to Vietnam, two weeks later, I had a letter waiting for me from Brockman. He thanked me profusely and said he would keep in touch.

True to his word, he sent me another letter some months later. But this time he said Lana was very worried about her mother and family and whether or not they even had enough money for food. The now-married Lana and her husband wanted to send her mother money. Lana's family was Roman Catholic, originally from North Vietnam, and she hoped the money would get there in time for Christmas. But in those days the Vietnam postal system was not reliable, so they asked me to deliver the money. I didn't have the heart to turn them down and agreed to help with the transfer. They wanted to give Lana's mother five-hundred US dollars cash in local currency, a small fortune in Vietnam in those days. This called for some careful planning since I figured the cash would fill at least a standard American grocery bag in piaster notes, the local currency. The inflation was atrocious.

I had an account with Bank of America from our previous Navy Base in San Francisco, and had used my account at the BOA branch in downtown Saigon to send money orders from time to time. I instructed the newlyweds to write me a check for the amount and send it to me through the Navy FPO mail system. When I received the check I deposited it in my account at the BOA in Saigon; then I waited two weeks for the check to clear. A few days later, my friend Dao and I got into a Navy jeep and drove downtown to the BOA. I put my .45 pistol under the seat for emergency. Dao waited in the jeep, I went into the bank dressed in my Navy khakis to look professional. I left the pistol in the jeep since it didn't seem right to take a loaded gun into a bank even if it was a war zone.

I approached a teller with my withdrawal slip and asked to withdraw $500 from my account in piasters. He gave me an odd look and said he would be back in a few minutes. I was invited to a room behind the teller's windows with some bank officers. They wanted to know what the cash was for, so I told them art pieces and paintings at an art dealer on Nguyen Hue Street. They asked more questions but that was my story and I was sticking to it. I felt a little uncomfortable when they asked if I had protection for this much cash. I said I'd be OK. I felt more uncomfortable when they brought me two large brown paper bags full of cash. The cash was in packets of bills bundled in banknote bands. There was an awkward moment when I needed assistance to put the second bag into my right arm. It surprised me how heavy they were and difficult to manage. Fortunately, they let me go out a private entrance close to where Dao had parked the jeep. I made it to the jeep, and Dao and I made it back to the office.

I hid the money in an old steel navy desk at the back of the office. The only way to contact Lana's mother was through Lana and Brockman, so I wrote them that I was ready for the transfer. Surprisingly the Navy FPO took only two days from Vietnam to Gulfport. They wrote back that they had set a time for the family to come to my place on Hai Ba Trung for the transfer. The date was to be December 23.

At 8:00 P.M. on the twenty-third I got a call from our Chinese guards at the front gate that there was a local Vietnamese boy to see me. I went to meet him but left the money hidden. It was dark at this time of evening and I didn't recognize him. He looked to be about thirteen, wearing shorts, no shoes and an old tee-shirt. He walked straight up to me and announced, "I'm here for money." I didn't feel comfortable turning over two bags of cash to a little kid in the dark. So I asked him in Vietnamese, "Can you return here tomorrow night at 8:00 with proof?" He shrugged, turned away and left.

Next day I had work at a project in Cu Chi, north of Saigon, and didn't make it back until nearly 7:00 P.M. I hurried through dinner and got ready for the meeting. I had taken the money and bundled it into a cloth bag that looked like the kind holding laundry, almost like a saddle bag. I kept the money hidden, but at 7:30 I went to a center area between our quarters and a warehouse facility for vehicles. This was originally an old French structure that had a large tiled courtyard with columns dispersed throughout that supported a tall roof. There was a high wrought iron gate across the front. I stood in the shadow of a column and waited for someone to show. From my vantage point I could see silver Christmas stars hanging from the Laotian Embassy across the street. I remembered it was Christmas Eve and my thoughts went back to a somber Christmas Eve the year before as I was leaving for Vietnam … it seemed a lifetime ago, and it occurred to me how unreal it felt that I would be returning home in only a couple of weeks.

At a little after 8:00 P.M. my thoughts were interrupted when I heard a motorcycle coming to the guard shack in front of the wrought iron fence. It was an old Peugeot model belching smoke as it came up the street. The same boy from the night before was driving, but this time an old woman was riding side-saddle on the back. She looked old and very small and frail and I didn't recognize her. She wore black sateen trousers and a well-worn long sleeve white blouse with large pockets on the sides like a man's suit jacket. I wondered if this could be Lana's mother. I didn't remember her this small. The street lights made her white blouse stand out against her dark face and trousers, and I couldn't see her face clearly. I stepped out to the street.

The young boy was on his best behavior. He shut down the motorcycle and greeted me with a cheerful smile and said his mother had come tonight. He said he was Lana's brother. I asked for proof and he gestured to the old woman. She slid down off the motorcycle and came around to me. I noticed she was barefoot like the boy. From below her trousers I caught a glimpse of her rough and calloused feet. Up close, I could see her deeply etched face. She reached into her right-hand blouse pocket and pulled out a crinkled-up piece of yellow legal paper. She smoothed out the paper and handed it to me. I looked at it under the light of the motorcycle head lamp and recognized it as the first letter I had sent to Gulfport before R&R. I told the boy, "You did good!"

When I retrieved the bag of money and presented it to them, the little boy was all smiles with excitement, but the old woman gasped at the size of the package and began to cry. The boy slung the bag over the back of the motorcycle and helped his mother onto the bike. Then he turned to me and said in Vietnamese, "Merry Christmas, Lieutenant." He started up the motorcycle, wheeled it around and headed back down the street, still belching a trail of white smoke.

★ ★ ★ ★ ★

I never saw or heard from any of those people after that Christmas Eve forty years ago, but still think of them now and then. I like to think they made it safely home and were able to put the money to good use. I like to think they were able to weather the changes in 1975 when the Communist regime took over. And I like to think they had a brief respite that Christmas, when the world was closing in around them, and a grateful and homesick daughter tried to make their lives a little brighter. I also like to think that little boy, now a man of perhaps fifty-some, still remembers driving through that night with a million dollars in his saddle bags.

How could he forget that?

Epilogue

On October 26 1972, Henry Kissinger announced, "Peace is at hand." It seemed the long Vietnam War was coming to an end, at least for the Americans. But the agreement fell through when all parties couldn't come together on terms, and the fighting and bombing continued for another three months, including the "Christmas bombing" when Nixon really poured it on. The Paris Peace Accords were finally signed on January 27, 1973. I said goodbye to my counterparts and friends on January 11, and left Vietnam on the twelfth, only two weeks before the agreement was signed. It was a truce to end hostilities, but most believed it would not really stop the fighting. There was too much bad blood between the North and South. The North was dedicated to their cause and had no reason to lay down arms after finally gaining the advantage as the Americans left. We were excited about the pending agreement nevertheless, because it would bring the release of some six-hundred American prisoners being held by the North Vietnamese. My replacement, already in the pipeline, showed up two days before I left and went home two days after me. It was the end of Vietnam for all but a handful of Americans.

On the morning of my departure I arose and dressed in civilian clothes. I was to fly to San Francisco and was told it best not to be in uniform, especially when landing in practically the epicenter of

On the morning of January 12, 1973, I dressed in civilian clothes and left Vietnam.

the hippie and anti-war movements. In those days, uniforms evoked in many Americans hostile feelings toward the military, everything from being baby killers to holding the military responsible for not winning the war. I must say, though, most citizens that I met in those days did not harbor ill will toward the returning veterans.

I landed at San Francisco International late afternoon after a stop of about one hour in Honolulu. Many on the flight that began in Vietnam changed planes in Honolulu for final destinations; only a few were left to fly on to San Francisco. When the plane touched down in San Francisco, those few produced a sparse but rousing applause in the big wide-bodied Pan Am 747 to mark the end of a long year. It did indeed feel good to be on US soil. I took a taxi to Treasure Island Naval facility in San Francisco Bay and booked a room at the BOQ. I called Linda from my room. It was a phone call long in coming, and one I'll never forget. I spent the next day mustering out of the Navy and was back in Ohio the following day. I was halfway around the world, out of Vietnam and out of the Navy, two days after I walked out of Nguyen Dinh Chu's office at the Saigon Shipyard. Two weeks later, during which Linda and I became re-acquainted while vacationing in Florida, I went back to work at General Motors.

I was still deeply tanned from a year in the sun when someone at work asked me during lunch, "Have you just returned from a cruise or something?" I answered that I was in Saigon until only weeks ago. He asked, "Oh yeah? Where's Saigon?" I think that's when reality hit. I soon learned that no one was interested in Vietnam. I was back in the USA. Vietnam became a taboo subject. The only ones I really discussed Vietnam with in any detail was Linda and my brother Jim who had been an infantry officer the year before I was there.

I was lucky in Vietnam. I was an advisor and didn't have to bear the burden of the day-to-day terror of military operations in places like the Central Highlands. My experience cannot be compared to those that did. I was fortunate to have a large degree of autonomy among the Vietnamese population and every day brought new experiences. I was never bored in that country. I have no bitterness about having spent a year in Vietnam, though I certainly understand those that do. Thousands were horribly maimed, physically and emotionally, thousands more paid the ultimate price. Nearly all were young persons who had a full measure of life in front of them.

Of the people I knew and worked with in these stories, I have lost track of virtually everyone. I think of Nguyen Dinh Chu sometimes and hope he had some good fortune and happiness in his life. Most were brave and good people living day-to-day as they could in an environment not of their choosing. I was anxious to get on with my life when I came home from Vietnam and was so focused on the future that I didn't take time to keep in contact with those I worked with so closely. I didn't realize then what an impact my time in Vietnam would have on me throughout my life.

A few days before my departure from Vietnam, Chu and I had a discussion about the future. I said, "Will we keep in contact?" He replied, "We may keep in contact, but I doubt it. Vietnam will change soon and it will be difficult to communicate. And besides we are both young with hopefully long and busy lives ahead. In Vietnam we have a saying, *"Xa mat, cach long."* Literally it means, when your face is far and I can't see you, your memory will gradually fade. He shortened it to, "Out of sight, out of mind." I said to him, "We must not let that happen." I'm sorry to say he was right and that it did happen. But Vietnam has always been on my mind.

Glossary of Terms

ARVN: Army of the Republic of Vietnam

Blue and Yellow taxi: Fleet of 1950s Fiat 500 automobiles that were painted blue and yellow and used by the public taxi company in Saigon

BOQ: Bachelor Officer Quarters; accommodations for US military officers for temporary stays

Cavitation: Destructive process in a centrifugal pump in which air is sucked into and implodes against the pump impellers

Central Highlands: Mountainous area in west central South Vietnam which extends to and past the western border with Cambodia. A sparsely populated and in many places heavily forested area that was the scene of heavy fighting throughout the war

Cholon: Chinese business district and neighborhood in Saigon; Chinatown, Saigon

CITROEN 2CV: Ubiquitous, small inexpensive French automobile developed in the 1950s and in production for decades, analogous to a Volkswagon Beetle. There were many imported to Vietnam especially during the 1950s when France considered Indochina a French Protectorate.

CIVIL ENGINEER CORPS: Officer Corps of the US Navy responsible for administering construction contracts, maintaining public works, and leading construction battalions.

CLAYMORE: Anti-personnel mine that sprays an arc of shrapnel triggered mechanically or by an electrical signal

CONCUSSION GRENADE: Grenade used to destroy enemy munitions and sappers under water

CV JOINT: Constant Velocity Joint; used to transmit power from vehicle axle to wheel hub.

DMZ: De-Militarized Zone; located at the seventeenth parallel of latitude between North and South Vietnam

DUC SIEU HOTEL: Civil Engineer Corps Officer Quarters in Saigon; was formerly a French Hotel prior to the Vietnam War

FPO: Fleet Post Office; for handling Navy mail worldwide

GI: Slang used to denote US military personnel, derived from the term "Government Issue"

Hooch: Temporary metal building designed to be erected quickly and used to house material or personnel. Slang military term for military barracks.

Jimmy 6-71: Shortened term for describing the General Motors Detroit Diesel engine comprised of six cylinders of seventy-one cubic inch displacement each

Lieutenant JG: Lieutenant Junior Grade (LTJG); second in level of rank in the Navy Officer Corps, equivalent to First Lieutenant in the Army or Air Force

Mamasan: GI slang term for Vietnamese woman or mother

M60: Machine gun of caliber 7.62 mm which fires the NATO cartridge and used extensively in Vietnam

M16: Standard service rifle of US military armed forces in Vietnam

Mike Boat: Small amphibious landing craft used by US military in Vietnam, formally Landing Craft Mechanized (LCM) with the nickname derived from the radio call letters Lima-Charlie-Mike

Mortar: Explosive shell which follows a natural trajectory after launch and explodes on impact

NIS: Navy Intelligence Service; that served in-country

North Vietnam: The portion of Vietnam north of the seventeenth parallel of latitude and officially termed The Peoples Democratic Republic of Vietnam

OICC: Officer In Charge of Construction; A US Navy Command term. In Vietnam, the Navy Department, specifically the Civil Engineer Corps, was delegated the responsibility of administering all construction contracts in-country.

PX: Post Exchange

Quang Tri: City of 30,000 just south of the DMZ which was attacked by North Vietnamese Army who crossed the DMZ on March 30, 1972 and which began what was termed the Eastertide Offensive

R&R: Rest and Recuperation; In general, military personnel serving a year-long term in Vietnam were permitted two weeks of time off, out of country or at designated R&R sites within Vietnam.

Seabee: Unofficial or slang term used to denote Navy Civil Engineer Corps Construction Battalions (CBs) as well as officer and enlisted personnel within that battalion.

South Vietnam: The portion of Vietnam south of the seventeenth parallel of latitude; officially The Republic of Vietnam

Tet: Festival celebrating the New Year in Vietnam, based on the Chinese Lunar Calendar

USN: US Navy

VC: Viet Cong; Revolutionary forces of the underground movement in South Vietnam under the umbrella of the National Liberation Front

VNN: Navy of the Republic of Vietnam

Warrant Officer: Naval Officer promoted through the enlisted ranks and receiving an officer commission of Chief Warrant Officer (CWO) based on performance of managerial and specialty skills.

WESTPAC kit: Slang term used by GIs in Vietnam to describe any number or collection of electronic and or camera equipment purchased at a discount available to service men and women in Vietnam through the US Military post exchange system.

6-71: Short designation for describing the General Motors Detroit Diesel engine comprised of six cylinders of seventy-one cubic inch displacement each.

green press INITIATIVE

The Wooster Book Company is committed to protecting the environment and to the responsible use of natural resources. The Wooster Book Company is a member of the Green Press Initiative, a nonprofit organization dedicated to supporting publishers in their efforts to reduce their use of fiber obtained from endangered forests. *Vietnam On My Mind* is printed on 30% post-consumer recycled paper that is manufactured through a chlorine-free process.

The Wooster Book Company works with printers such as Thompson-Shore, Inc. that are also members of the Green Press Initiative and who help establish the chain-of-custody for this initiative by supporting best practices developed by the Forest Stewardship Council.